S0-ARY-495

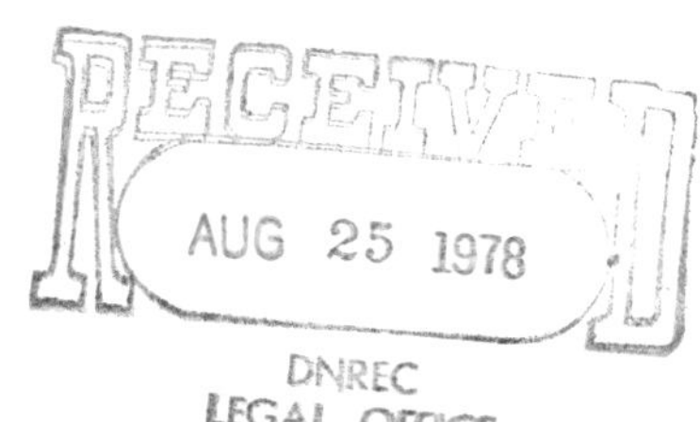
RECEIVED
AUG 25 1978

Ocean Thermal Energy Conversion

Ocean Thermal Energy Conversion

Legal, Political and Institutional Aspects

Edited by

H. Gary Knight
J.D. Nyhart
Robert E. Stein

Published under the Auspices of
The American Society of
International Law

Lexington Books
D.C. Heath and Company
Lexington, Massachusetts
Toronto

Library of Congress Cataloging in Publication Data

Main entry under title:
Ocean thermal energy conversion.

Includes bibliographical references.
1. Ocean thermal power plants–Law and legislation. 2. Ocean thermal power plants–Law and legislation–United States. 3. Maritime law. 4. Ocean thermal power plants. I. American Society of International Law.
K3984.025 341.7'55 77-2049
ISBN 0-669-01441-9

 This book was prepared with the support of NSF Grant AER 75-00280 A02. However, any opinions, findings, conclusions or recommendations herein are those of the Authors and do not necessarily reflect the views of the National Science Foundation.

Published simultaneously in Canada.

Printed in the United States of America.

International Standard Book Number: 0-669-01441-9

Library of Congress Catalog Card Number: 77-2049

Contents

Introduction

The sea has always sparked human imagination. A fairly recent example is the concept of ocean thermal energy conversion (OTEC)—capturing the sun's warmth in the ocean surfaces and harnessing the temperature differences of surface and cold and deep waters to generate energy. The idea—a nonpolluting energy source, endlessly rekindled from the sun, at costs comparable to current energy sources—over two generations has successively caught the imagination of a handful of scientists, government energy programmers, and most recently industry innovators. To build entirely new seagoing structures that will house heat exchangers yet to be perfected in order to coax from the very elements of the seas' waters a greatly needed new resource—these challenges engage not just the visionary, but also the hard-headed wielders of technology.

Technically, the theoretical basis of OTEC is simple. Warm seawater from the ocean's surface and the cold deep water below are pumped through a heat exchanger that employs a working fluid, such as ammonia, propane, or Freon, in a classical closed cycle. The warm water vaporizes the working fluid which turns a turbine.

In theory the electricity generated by the turbine can be used in a number of ways. It can be transmitted via cable from the offshore plant to shore and integrated with an existing energy grid. It can be used on site in energy-producing processes such as uranium enrichment or hydrogen production for subsequent use elsewhere. Or the energy can be used on site to refine metals such as in electrolytic reduction for aluminum, titanium, lithium, etc.; for ammonia production for fertilizer; and for other high-energy products. Current ERDA plans project a 25-MW prototype plant by 1985, a 100-MW demonstration by 1988, and the possibility of beginning commercial production two years later. Other groups have projected smaller but earlier use of OTEC, coupled with onsite production.

If OTEC succeeds technically and economically, there will be extensive legal questions associated with it. The time to recognize and plan for this possibility, however, is well in advance of any demonstration or prototype trials. It was for this reason that the American Society of International Law began the task of investigating aspects of the future legal framework for OTEC. Significant technological innovations inevitably evoke their own response from the law. Too frequently the reaction comes only after a passage of time during which one or more problems arise from the new use to which public concern is directed. Part of the lawyer's job is to help anticipate such problems and structure a legal framework for avoiding, minimizing, and, most of all, resolving such problems when they do come up. OTEC challenges the lawyers and policymakers in this task just as it challenges the engineers in theirs. The panel, whose work forms the body of this book, was established to examine these issues.

The panel's study, however, does not evaluate the technical feasibility of OTEC. Considered opinions may be found on both sides of the issue, and it is not the function of this study to take a position. Rather, we have assumed that at some point the existing technological and economic challenges will be met, while recognizing that in fact this might not be the case.

Neither was it necessary to the work of the panel to know now the details of the technology yet to be worked out. Accustomed to dealing with hypothetical questions the lawyers, political scientists, economists, and marine specialists working on the panel are comfortable in asking "what if?" and proceeding from there. What would be the legal implications and issues if an OTEC operation were mounted in the near future? This line of questioning is feasible because any new technological innovation comes into being surrounded by an already existing legal environment. The problems and prospects arising from the existing legal framework form the backbone of this study.

While the technological and economic problems are being worked through, it is critical that those solving them be aware of the attendant legal issues. The fact that this study was commissioned to examine the legal issues at this early stage in the development of OTEC is very unusual in the history of federal support for technological innovation. The pioneering concept of including the legal issues among the technical, biological, and economic concerns now covered in the ERDA and NSF programs of OTEC development bears witness to the farsightedness of both agencies and of the persons in those institutions responsible for the programs.

Study of the legal aspects of OTEC holds further significance. OTEC's intrinsic need to utilize ocean space adds to its prototypical character. Technological capacity to use the oceans in new ways has proceeded with the unanticipated pace that typifies so much of technology in modern society. Other new forms of energy production are at the preproduction or experimental stage. Technical capability in oil and gas has moved rapidly into deeper waters and larger operations, although today there is at least a pause in the forward rush of development which serves as a reminder that, ultimately, economics still binds technology.

Characteristics which OTEC shares with other future uses of the ocean give rise to serious, and common, legal considerations: its purpose is not seabed exploitation and exploration and therefore it is not covered under the 1958 Continental Shelf Convention; it is located outside existing territorial waters, yet inside a prospective economic zone or extended jurisdiction; there are possible physical links to coastal states; U.S. and state/city legal implications might apply; there is no specific legislation or treaty; it has questionable status as a vessel; there may be a time delay pending operational status during which the law of the sea will change.

These commonalities lend wide significance to the study of the legal considerations of OTEC. Many of these considerations have been investigated by panels and working groups established by the American Society of International Law (ASIL) which have dealt with environmental protection of the oceans, problems associated with the exploitation of the deep seabed and the living resources of the sea, and a variety of technical law-of-the-sea issues.

Both OTEC's potential contribution to the world's energy supply and the legal/technical issues involved, which will have broad significance independent of OTEC's future, have motivated the American Society of International Law panel throughout its study. The panel's objective has been to consider the legal and policy implications of the development of OTEC technology under present and prospective domestic and international law. Thus the panel is one part of the ASIL's continuing program of interdisciplinary study which probes a wide range of international policy issues where law has an important role and whose examination is required by scholars and government officials.

The chapters can be grouped into three headings: nonlegal issues, international legal aspects, and domestic regulation.

The first group provides an overview to nonlegal issues. Sheets's study (Chapter 1) provides the technical background essential to an understanding of the legal and policy aspects. It reviews and summarizes the very broad range of technical concepts which go into making an OTEC plant feasible. Sheets captures well the crucial aspect of the issues that remain to be faced. Stern's treatment of the economic prospects of OTEC (Chapter 2) reminds us all that the ultimate trial of OTEC can be expressed as an economic one, in which this enticing energy source must meet satisfactorily all the criteria—legal, technical, political, and financial—discussed in this book.

The second group—Chapters 3 through 7—concentrates primarily on the international legal aspects.

The Third United Nations Law of the Sea Conference may well rewrite the familiar jurisdictional boundaries of the oceans, foreshadow new ocean uses through the creation of coastal state economic zones, create new environmental standards and duties, and set out rights and responsibilities regarding ocean research and transfers of technology. Even if the negotiations are not successful, states may well adopt some of these new concepts through domestic and regional arrangements. All would have an impact on proposed OTEC projects, located offshore near either developed or developing countries. Moreover, if there is no treaty, alternative means will be developed for meeting national policy objectives, such as increased reliance on new ocean uses that are acquiring legitimacy through customary law.

Hollick's study (Chapter 4) sets the broad political and geopolitical framework, while in Chapter 3 Knight uses as a departure point the jurisdictional issues which must be solved as a first part of the international puzzle laid out by Hollick.

The extent to which an OTEC operation, whether one platform or a whole network, displaces or constrains other uses—navigation, fishing, oil and gas exploitation, mining, recreation, and military or industrial purposes—will bear upon the operation's standing in international law as a reasonable use of ocean space. Washom examines these issues in Chapter 5. Thus we come back to Hollick's consideration of the political environment in which OTEC will be launched.

Future changes in international law must be set in a context of already existing international regulatory agreements, standards, and requirements which would fall on an OTEC operation today. In turn, these lead to examination of the kinds of international regulatory framework which might be envisaged over OTEC operations. These considerations are expanded in Chapter 6 by Hallberg.

The planning and operation of an OTEC facility and associated onshore and offshore installations create a number of potential environmental problems. They may cause changes in the oceans' waters and perhaps in the atmosphere which can have an impact far removed from the site. While other ERDA projects are examining a variety of OTEC's possible environmental effects, many of the legal issues raised are predictable even in the absence of much needed data. The international aspects of these questions are considered by Stein in Chapter 7.

The third group—Chapters 8 through 11—concerns U.S. domestic regulation and conflict management.

The impact of maritime law on OTEC operations bridges both the international and the national sectors. The adaptability of traditional maritime law to new uses of the ocean is the concern of the first half of Nyhart's work (Chapter 8). Subsequently Nyhart turns to the private law questions of liability and responsibility which might arise from OTEC operations and which will be handled largely as either maritime or domestic issues.

Alongside private law issues lie the concerns of Higgins (Chapter 9), that is, the domestic regulatory scheme that would oversee an OTEC operation. The U.S. Coast Guard, Corps of Engineers, Environmental Protection Agency, and the Federal Power Commission as well as other federal and state agencies would all have extensive regulatory reach over any OTEC over which the United States claimed jurisdiction.

In a somewhat similar vein, in Chapter 10 Stoel deals with the environmental regulations and other legal concerns through the perspective of U.S. federal and state law.

Finally, the financing of an OTEC operation in large part may determine its organizational character. The major determinant for whether U.S. OTEC operations will be publicly owned, privately owned, or a mixture may well be that form of ownership which can bring together the substantial necessary funding. In Chapter 11 Riggs reviews financing methods, drawing on existing experience, particularly in the offshore oil and gas drilling areas.

In order to provide some consistency through the study, a small number of common assumptions have been made. Unless otherwise indicated, we assume that OTEC is operating somewhere off the continental United States plus

Hawaii, at a distance of between 3 and 200 mi. In a number of chapters consideration is given to the implications of OTEC operating on the high seas, either beyond what will likely evolve as coastal economic zones, that is, 200 mi, or within the economic zone of another country. These assumptions are stated explicitly. There are two assumptions about the connection to land: (1) the OTEC would be linked to a land-based power grid by a cable, and (2) its energy would be taken off in another form or used at the structure. Finally, unless it is otherwise indicated, we have assumed that OTEC is moored, although it may occasionally use on-board thrusters to provide dynamic positioning assistance.

As is indicated below, the study took place in about one year. A number of issues were identified which could not adequately be treated.

Policymakers, most immediately those in the U.S. government, have a considerable stake in the wise and orderly evolution of the legal frameworks within which offshore uses develop. Therefore, it seemed appropriate to note briefly some of the areas of future legal and policy development suggested by this study. They fall into three broad categories.

First, there is the general issue of the law for offshore structures and uses as a generic class. The prototypical character of OTEC has been noted above. To date, the development of offshore uses which depart from traditional arenas, i.e., transportation and fishing, has been treated from a legal viewpoint in an ad hoc manner. This has been true both within the United States and in the international community. That OTEC raises issues for a whole class of vessels suggests that an early order of legislative business would be to examine this range of uses, to identify the legal and regulatory needs common to all, and to address them uniformly. Differences in the applicability of maritime law and ambiguities involving structure classification as vessels or fixed installations exist and should be ironed out. The division of environmental responsibility between federal and state governments should be clarified. Lead agency assignments in various areas should be specified. Other needs that apply across the board to offshore developments are indicated in various chapters and the Appendix. The point is that a clear legislative approach is needed for the development of offshore uses and structures generally.

The second area, the need for legislation specifically applicable to OTEC, can be divided into two components. First are the issues applicable to the initial development and promotion of OTEC. If we assume that Congress and the Executive Branch will settle on a program and space of development, several questions arise that require simultaneous answers. The movement from idea to reality will require an institutional home for OTEC. Studies will have to be made in order to decide whether to make use of existing agencies, to develop a new agency, or to employ some combination to house both the promotional and the regulatory aspects of the new technology. In large measure this will depend on whether the development of a prototype and a system is largely governmental, shared between government and industry, or entirely industrial. As a part of the promotional stage, the question of governmental assumption of financial and "disaster" environmental liability ought to be considered. For example, will a new "Price Anderson Bill" be needed, or will an international oil pollution fund be adopted? Other legal matters attend the promotional stage. An environmental impact statement required by NEPA will have to be issued. Further, internationalizing the NEPA process may lead to a responsibility on the part of the United States to inform other countries of the proposed environmental impacts of the technology at that time.

If we assume that the developmental effort proceeds successfully, then the commercial role of OTEC in the energy order will have to be discerned and its role determined from a legal and institutional perspective as well. If OTEC does become a reality, consideration will have to be given to the type of legislation needed to govern liability and responsibility for the technology itself. The commercialization of OTEC will require analysis of the kind of rate structure to be used for the electricity produced.

Finally, the third area has both international and domestic aspects. If the technology developed is one which produces ammonia or hydrogen on board and then tanks it in to shore, what sort of tanker standards will be required? Can existing technology be used, or will there be a need for new standards? Will these be domestic or international, perhaps worked through the Intergovernmental Maritime Consultative Organization (IMCO)?

We continue to believe that the legal assessment and development work must keep abreast of the technical if the beginning taken by the assignment of the chapters that follow is to be maintained.

The OTEC panel was established.as a result of a grant to the ASIL by the Research Applied to National Needs Directorate of the National Science Foundation (NSF). Shortly after the grant began, technical aspects of the ocean thermal program were moved to the Energy Research and Development Agency. While this effort remained at the NSF, the panel also benefited considerably from the participation of ERDA representatives.

It has proved to be a successful ASIL practice to develop membership in its panels from a variety of sources. So, with the OTEC panel, a broad base of technical, economic, and political expertise, in addition to legal expertise, was sought. We feel that the group selected enabled the panel to obtain a broad and balanced view of both the technology and its legal, political, and institutional implications. The panel held five one- and two-day meetings from February 21, 1975 to December 4, 1975. At many of these meetings the panel was joined by a number of guests. These included practicing attorneys working with OTEC grantees and representatives from both Lockheed and TRW who had carried out studies for ERDA on the development of an OTEC model. These individuals provided helpful insights into a range of issues and practical problems. A List of Contributors can be found at the back of the book.

In order to present the findings of the panel to a broader audience, ASIL organized a two-day workshop, also sponsored by the National Science Foundation and the Energy Research Development Administration. Held at the Mayflower Hotel in Washington, D.C., on January 15 and 16, 1976, the workshop brought together more than a hundred participants from the legal and academic communities, industry, and government. The report of the workshop is included as an Appendix.

The work of the panel benefited as well from the participation of Dr. Robert Cohen, Program Officer at ERDA; Dr. Arthur Konopka and Norman Wulf of the National Science Foundation; and John Lawrence Hargrove, Director of Studies at ASIL. Eva Sheldon and Gerald Fisher of the ASIL staff also assisted the panel. Valerie Hood and Judith Hall are to be thanked for their assistance in preparing for the workshop. In many ways it was a community effort. But, while the panel benefited greatly from all the assistance it received, this book is the responsibility of the authors; its conclusions, findings, opinions, or recommendations do not necessarily reflect the views of either the sponsoring institutions or the organizations from which the panel members and guests were drawn. Nor does it constitute the views of the American Society of International Law, which as an organization does not take formal positions on matters of the kind dealt with in this study.

Robert E. Stein
Principal Investigator

H. Gary Knight
J.D. Nyhart
Panel Cochairmen

1 Ocean Thermal Energy Conversion (OTEC) Plants: Technical Background

Herman E. Sheets

Introduction

This chapter serves as the technical background for the Panel on Ocean Thermal Energy Conversion of the American Society of International Law (ASIL). It summarizes in broad terms both the work done to date and the findings of the various research teams. The chapter will also emphasize those systems and components that require extensive development and constitute high-risk items.

Ocean thermal energy conversion—OTEC—has been considered for some time. In 1881 Jacques D'Arsonval proposed the concept of using the temperature difference in the ocean as a source of energy. In 1930 Georges Claude attempted to build an OTEC power plant off the coast of Cuba. Only Claude's open steam cycle project reached the hardware stage, and none of these proposed projects was successful. More recently, a substantial effort is being made in the analysis of various OTEC concepts. Published data include the MacArthur Workshop on Energy from the Florida Current and a substantial number of papers in professional engineering journals.

Basically, there is a potential for a substantial amount of energy due to the temperature difference of water at the upper and lower levels of the ocean. The exploitation of this thermal energy is intriguing, provided it can be harnessed economically for use near centers of energy consumption. The maximum temperature difference in ocean waters occurs near the equatorial regions. Geographic data indicate that these locations are not located close to the centers of energy consumption. If OTEC is practical at less than the maximum temperature difference, then the geographic area of potential locations of power plants is substantially increased. Certain locations near the coast of Florida and off the coast of Hawaii have substantial differences in ocean water temperature which could justify OTEC plants provided they were technically and economically feasible.

Five major research teams have been working on the OTEC concepts under National Science Foundation sponsorship for some time. Each team has proposed a different configuration, and there are three different fluids for the proposed systems. All systems are using a closed cycle, and they require warm- and cold-water circulation systems with energy transferred through heat exchangers. The major teams are as follows:

1. Carnegie-Mellon University (CMU)
2. University of Massachusetts (UMass)

3. Sea Solar Power, Inc. (SSPI)

These organizations have contributed the original research, and two industrial organizations have made a technical and economic assessment:

4. Lockheed Missiles and Space Company
5. TRW Systems Group

Since all systems use the closed-cycle power process, a brief description of it follows. The ocean acts as a collector of solar energy, and ocean thermal energy is a form of solar energy. The heat from the solar energy increases the temperature near the ocean surface, and the water temperature in the depths of the ocean is substantially lower. Thus, the ocean becomes a natural energy collection-and-storage device. The OTEC plant converts thermal energy into mechanical energy and eventually into electric energy. The thermal-to-electric conversion is accomplished in a cycle using a working fluid such as propane or ammonia. Figure 1-1 shows the basic principle. Warm water enters the OTEC

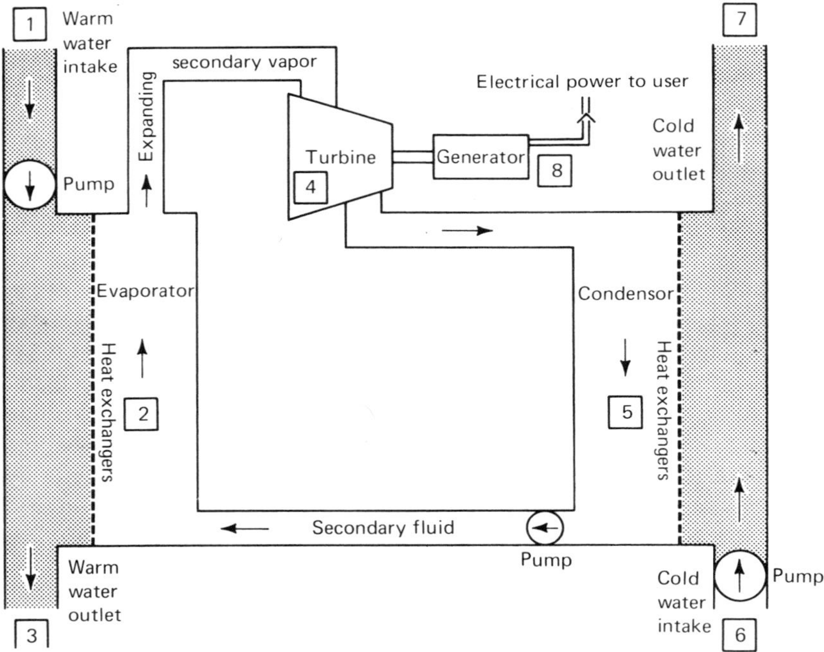

Figure 1-1. Schematic Diagram of a Closed-Cycle Ocean Thermal Power Plant.

plant at location 1, is pumped through the heat exchanger at location 2, and leaves the plant at location 3. The heat exchanger-evaporator (2) vaporizes the working fluid, for instance, ammonia. This vapor is expanded in a turbine (4); then it leaves the turbine to enter the condenser (5). From there a pump returns the secondary fluid to the heat exchanger-evaporator (2). The cold water enters at location 6 and flows through the heat exchanger-condenser (5), leaving the plant at location 7. The turbine operates the electric generator (8), providing electric power to the user.

Note that the temperature difference between the hot and cold water is approximately 40°F, which results in a very low theoretical Carnot efficiency of the thermodynamic cycle, namely about 6 percent. The achievable net efficiency of an OTEC plant is expected to be about half this theoretical value. Thus, the conversion efficiency of this plant is quite low compared to the 30 to 40 percent net efficiency of large fossil-fuel thermal power plants. Therefore, an OTEC plant must circulate large amounts of warm and cold water to produce energy. However, remember that no fuel is required, and the amount of available heat energy is very large and practically inexhaustible. An OTEC system operating at sea will be contained on a platform, and a number of configurations have been proposed. Those plants operating near shore can deliver electric energy directly to the users on land. For more remote locations, it has been suggested that intermediate products such as hydrogen, methanol, ammonia, or other high-energy products be manufactured at the OTEC site. An OTEC plant does not require new technology; it needs to refine and optimize existing technology in many areas. In addition, certain potential environmental and legal questions must be solved before such a power plant can be operated at sea. The following chapters will describe the state of the present investigations.

This chapter reports the findings of the above organizations in general terms, and it will identify critical issues affecting technical and economic viability of commercial implementation of OTEC systems. The chapter will also evaluate those components and systems which have relatively high risk and thus can affect the operation of the entire power plant. The above-mentioned projects and this chapter review only ocean thermal energy and the thermal power plant, including the electric generator. No consideration is given to transmitting electric power from the OTEC plant to the user or making the OTEC power plant part of a utility network. If the electric output of the power plant is not too large, on the order of 500 megawatts, and not too far offshore, then existing electric cable technology can provide for power transmission. For large power and long distance, power transmission can present significant technical and economic problems.

Basic Technical Information

Basic Work on Ocean Thermal Power Plants

The basic work on ocean thermal power plants consists of studies by three teams:

1. Carnegie-Mellon University (CMU)
2. University of Massachusetts (UMass)
3. Sea Solar Power, Inc. (SSPI)

The CMU concept consists essentially of a spar buoy with a large, vertical inlet pipe. The warm-water intake is at the top near the ocean surface, and the warm-water discharge is at a lower level near the boiler modules. The cold-water inlet is at the bottom; the cold-water discharge is near the condenser modules at a higher elevation, a considerable distance from the cold-water inlet. The UMass design consists of twin horizontal hulls somewhat akin to submarine structures. The warm-water inlet is above the cylindrical hull, whereas cold water is pumped from a lower level up to the altitude of the cylindrical hulls. Note that this configuration does have structures penetrating the surface of the water, and therefore it has the capability of access during operations. The SSPI concept consists of a floating platform. In this case, the condenser is above the boilers, thus eliminating the boiler feed pump.

The published information shows different degrees of completeness and covers some areas in considerable depth while other areas are not treated with equal thoroughness. Basically, the main emphasis in previous work has been on the power plant cycle—the turbine machinery together with the heat exchangers. The ocean platform and its operating requirements have been analyzed in terms of basic configuration but have not been treated in substantial depth. Admittedly the technical complexities of the platform are of smaller magnitude than those of some of the other subsystems. However, its configuration, construction, and method of deployment can substantially affect the total cost of the system. The three research teams have also made cost and performance estimates of their proposed systems. However, the dates and assumptions on which these various estimates were made have not been uniform. As a result, the National Science Foundation has issued two contracts to industrial organizations, TRW and Lockheed, to evaluate the studies of the three teams and establish a performance and cost basis for an OTEC power plant. In the process of this evaluation, Lockheed has established a baseline design, whereas TRW made its evaluation on the basis of the work of the three teams.

The Lockheed baseline design consists of a vertical spar-type buoy extending a considerable distance downward, anchored to the sea floor. The platforms must be kept at a fixed position or at a predetermined location if electric power is transmitted by cable. However, if energy-intensive products are developed on site, a fixed position may not be required from an engineering point of view, and then movement can be considered.

Plant Size

Each of the above plants operates on a somewhat different concept and has a different power output. Table 1-1 shows the various gross and net power outputs

Table 1-1
Power Output of OTEC Power Plants

Proponent	*Gross Power Output, MW*	*Net Power Output, MW*	*Power Output Ratio*	*Module Size Net, MW*
CMU	140	100	1.40	25
UMass	500	400	1.25	25
SSPI	125	100	1.25	25
Lockheed	250	160	1.56	40

and the module size for each of the proponents. The Lockheed baseline design is also shown in the table. There is no uniformity in power output, module size, or ratio of gross to net power. The data from TRW and Lockheed are more conservative than those of the three original proponents, and this is also indicated by Lockheed's value of 1.56 for the ratio of gross to net power. There are also differences in the assumption of both evaporator and condenser entrance temperatures. The assumptions regarding the hot-water temperatures are as follows:

CMU	80°F
UMass	72°F
SSPI	76°F
Lockheed	80°F
TRW	79°F

Condenser entrance temperature is 40°F for all cases, except Lockheed which has a variable condenser temperature with values of approximately 47°F. The warm-water and cold-water temperatures are a function of the selected site.

The various proponents have assumed different heat-transfer coefficients. The heat-transfer coefficients based on the various concepts are as follows:

U = 200 to 400 Btu/h · ft^2 · °F	UMass, SSPI
U = 1000 Btu/h · ft^2 ·°F or larger	CMU
U = 500 to 1000 Btu/h · ft^2 · °F	Lockheed
U = 415 Btu/h · ft^2 · °F	TRW

The wide range in heat-transfer coefficients is based on different rates of water flow on the water side and different concepts of heat-transfer enhancement by the engineers of the proponents. The assumed heat-transfer coefficients are above those used in present seawater condenser practice. There is presently no equivalent for the seawater-evaporator heat exchanger. However, this heat exchanger would be quite similar to the condenser except for the temperature level.

There is less difference between the proponents regarding the assumptions for the turbine, generator, and the pumps. However, each team uses a different fluid. CMU uses ammonia, UMass uses propane, and SSPI uses a Freon-type refrigerant. Lockheed prefers ammonia but is also giving consideration to propane.

Time Frame for Ocean Thermal Power Plants

Each group has looked at the time frame for OTEC and has considered it essentially a long-term project. As a result, there is no clear indication when the first unit might be expected to be in service or what power it might produce. The Lockheed time frame extends to the year 2000, and it is assumed that a prototype OTEC power plant will be ready for production by 1985, with substantial power being generated in the year 2000. An experimental prototype could be available in 1980 if appropriate funding becomes available. The TRW report shows a time estimate for CMU and for UMass, both indicating a conservative approach beginning design of a production plant in about 8 or 12 years, respectively. Building such a production plant has been estimated to require an additional 3 years, so that the entire project may require 11 to 15 years after initiation. Lockheed's estimate indicates that the total time frame for a large plant would require approximately 15 to 20 years after initiation. It indicates that production plants could be online by 1990.

The magnitude of these projects together with the substantial length of lead time has led the various investigators to make estimates about the potential improvements and developments which can be expected during the next 20 years. Such projections involve a substantial risk since technological improvements depend to a considerable extent on investment in capital and resources, which in turn will be affected by policy and general economic conditions. Each investigator has his or her own concept regarding possible technical and manufacturing improvements; therefore, each report reaches a different conclusion regarding performance and costs. It must also be recognized that most engineering projects begin with a small prototype and the estimates for production units are not made until after a prototype has operated successfully. No doubt this can be done for OTEC power plants, but at this time no prototypes are in operation. Some of the investigators compare the cost estimates for large OTEC units with fossil-fuel or nuclear plants. However, this comparison can be meaningful only if some reasonable assumptions can be made regarding the fuel costs and technical development for these plants during the next 20 years.

Technical Subsystems

Heat Exchanger

Design. The heat exchanger is by far the most important component of the ocean thermal power plant in terms of size, cost, and performance. If the heat exchanger does not meet the expected performance, the entire thermal cycle could be downgraded so that the amount of power which can be generated becomes substantially lower. The heat exchanger is also the largest component in both physical size and cost; therefore, it becomes an all-important subsystem to make OTEC a practical solution for the generation of power. The performance, size, and cost of a heat exchanger are greatly affected by the heat-transfer coefficient which is built into the unit as part of its design.

For the lowest heat-transfer coefficient, the heat-transfer area is about 7 times as large as for the highest heat-transfer coefficient. Obviously, for the ocean thermal power plant, various designs of heat exchangers have been proposed which plan to achieve a higher heat-transfer coefficient than is presently in use with seawater condensers and with other components in ocean use. Some research has been done in which higher heat-transfer coefficients were attained in the laboratory for certain fluids and materials. Additional development work is needed in the area of tooling, manufacturing, and operating as well as design before a heat exchanger with substantially improved performance can be used in the ocean. The heat-transfer coefficient also affects critically the cost of the heat exchanger. There is presently some uncertainty regarding the cost per square foot of heat-exchanger area.

The design of a heat exchanger is affected by the available temperature difference and the amount of water flow through the heat exchanger. High water velocities improve the heat-transfer coefficient, but they require larger pump capacity and horsepower. On the other hand, a larger temperature difference is conducive to transferring larger amounts of heat through the same area with the same heat-transfer coefficient. Therefore, a higher temperature difference could result in lower ocean water velocities under otherwise identical conditions. Under otherwise identical conditions there will be a reduction in ocean water flow (in gallons per minute) with increasing temperature difference. The water volume required for the ocean thermal power plant is very large; thus pumps of extraordinarily large size and substantial power consumption are needed.

Fluids. Each of the three teams is using a different fluid: CMU is using ammonia, UMass is using propane, and SSPI is using Freon. There are also different concepts regarding the arrangement of the heat exchangers. SSPI eliminates the boiler feed pump by locating the boiler below the condenser. The physical dimensions are dependent on the specific gravity of the fluid used (Freon in this case). Other designers prefer to arrange boilers and condensers at a water level in the ocean where the vapor pressure of the fluid used in the power plant system is equal to the ocean water pressure; thus the pressure differential across the heat exchanger at a minimum. This can be done in both the boiler and the condenser at the expense of some other compromises.

Other criteria for the heat exchangers are that they should be reasonably accessible for underwater maintenance. This makes it desirable to keep submergence at less than 300 ft and preferably less than 20 ft, thus controlling the water pressure and the fluid vapor pressure. This factor will indirectly influence which fluid is used in the power plant cycle. Another consideration is the total price of the working fluid, since in an ocean thermal power plant the fluid inventory is of considerable magnitude. Finally, the selection of the fluid will also require evaluation of toxicity, as well as fire and explosive hazards, together with operational factors. Ammonia has some problems in several of these areas; propane has fewer problems, and Freon may be best from a safety and handling point of view. On the other hand, Freon is the most expensive of the fluids, and ammonia would result in desirable higher operating speeds of the turbine and thus lower turbine costs.

Choice of Materials. No clear selection of the material for the heat exchanger has been made. Aluminum has been suggested as a heat-exchanger material, and among the many choices it has the advantage of lower initial costs and good heat-transfer capabilities. From the point of view of maintenance, a tubular heat exchanger with seawater flowing inside the tube appears best because it can be kept clean by a mechanical cleaning system. The shell and tube heat exchanger also has advantages because of its rugged structure and handling. However, the tube-and-shell type of heat exchanger may be larger and heavier. The extended-surface heat exchanger can be made rugged and compatible with the sea environment. However, at this time, the extended-surface type of heat exchanger has not been built in the size required for an OTEC power plant. Even in smaller sizes, the present methods of construction are relatively costly.

Other materials which have been suggested for the heat exchanger are 90-10 Cu-Ni (copper-nickel). Presently this material is used extensively for condensers and feed water heaters. It performs well, but its cost is high. Substantial experience with it does exist for the shell-and-tube type of heat exchanger in seawater. Another material which has been recommended is titanium. Titanium has been used to a limited extent in seawater. It has excellent corrosion characteristics, but its price is considerably above that of copper-nickel and manufacturing experience is limited.

A plastic material has also been proposed for use in heat exchangers. It consists of a high-density polyethylene plastic, possibly loaded with up to 30 percent aluminum powder. Practical experience with this type of material is extremely limited, and it may prove usable only in small wall thicknesses.

The heat exchanger and its materials have been identified as the most costly element in an OTEC power plant. Therefore, considerable effort has been made to optimize heat-exchanger design and to improve the heat-transfer coefficient in order to reduce the amount of material used. As a result, various designs have proposed internal or external devices to increase the heat-transfer coefficient. Such features will require new developments and may be justified if there is a large demand for heat exchangers for OTEC power plants. However, the amount of improvement in heat transfer and consequently the reduction in material can be ascertained only after a test program has been completed. This will also require the development of manufacturing methods so that heat exchangers can be built from selected materials on an economic basis.

Performance. The heat exchangers represent the major cost element in an OTEC system because of their size. At this time, no final selection has been made regarding the material, design, or manufacturing methods. In a salt-water environment, 90-10 Cu-Ni or titanium may be preferred. As to the internal fluids, propane is the most compatible with any of the proposed heat exchanger materials. Ammonia has the best heat-transfer properties, but small amounts of moisture can make the fluid highly corrosive. Seawater contains salt in solution and organic matter that will produce fouling and scaling which, in turn, reduce the heat-transfer efficiency of the heat exchanger. The investigative teams have put different emphasis on the various heat-exchanger characteristics. One team considers a high overall heat-transfer coefficient of greatest importance. Another team places greater emphasis on the prevention of biofouling, and yet a third team emphasizes light weight and low cost as the highest-priority item.

In summary, the successful OTEC power plant requires improved heat transfer technology. If titanium or copper-nickel is selected, the large weight of the heat exchanger could absorb nearly the entire supply of these materials in this country. For both the evaporator and condenser, a single OTEC plant requires between 4 and 10 million lbs of material. The heat exchanger and its material are the most critical element in the construction of an OTEC power plant.

Power Plant

Thermodynamic Cycle. All three teams use a Rankine cycle as the most attractive for converting the available temperature differences in the ocean into usable energy. The Rankine cycle produces vapor which is expanded in a turbine and subsequently condensed. The condensed fluid is returned to the evaporator to form a closed cycle. There is no clear indication as to the preferred working fluid. Ammonia has the lowest molecular weight, has a relatively high volume, and permits a high turbine speed. Freon has the highest molecular weight, a relatively low volume, and a low turbine speed. It also requires the largest amount of mass flow and, therefore, a considerable inventory of fluid in the system. Propane has characteristics between the above two extremes. Due to the small temperature differences, the theoretical cycle efficiency is relatively low, and the theoretical Rankine cycle efficiency is very close to the ideal Carnot efficiency. The heat absorption and rejection rates are essentially independent of the working fluid. The volume flow is lowest for ammonia, higher for propane, and highest for Freon. It is noted that the total efficiency is on the order of 3 percent for the ratio of power delivered at the shaft to the heat absorbed from the ocean. The total plant efficiency as the ratio of net electric output to heat absorbed is approximately 2.3 percent. The uncertainty over achievable heat-transfer coefficients can distinctly affect plant efficiency; and if the parasitic loads required for pumps and other auxiliaries are higher than estimated, the total power output may be substantially reduced.

Turbines. The turbine output must be on the order of one-third larger than required for electric power generation in order to provide the power for the auxiliary components. The proposed turbines can be manufactured with available technology and can be expected to have high efficiency. The selected power for the modules of 25 MW or larger appears reasonable. However, at present such turbines are not commercially available, although similar units have been in use in the chemical industry. Ammonia requires the smallest turbine diameter and the highest shaft speed, whereas Freon requires the largest turbine diameter operating at the lowest speed. Careful attention must be given to the selection of bearings and the method of lubrication for the three recommended fluids. The three teams have all assumed a 90 percent turbine efficiency. While such an efficiency is possible, it may require some developmental work. The turbine also requires careful material analysis, particularly in the case of ammonia, as it requires the fastest-running turbine.

Controls. Instrumentation and controls so far have received little attention. The standard power plant instrumentation will be required to monitor and control the entire system. The ocean environment will impose additional requirements to ensure safe operation for personnel and equipment in the ocean environment. The CMU concept envisions an unstaffed, fully automatic power plant. This is an innovation which imposes additional requirements such as a telemetering system to monitor all operations by remote control.

Present power plants have control systems where the energy input or fuel into the power plant is controlled relative to the energy demand by the system. This is true for both nuclear and fossil-fuel plants. For the OTEC power plant, however, the amount of available energy changes as a function of many factors, and the available energy is transmitted to the turbine and the generator. It may require an additional control to match power generation to the power demand of the system. Instrumentation will be required to monitor power output, voltage, frequency, and temperatures at a number of locations in the heat exchangers and in the warm- and cold-water system. The composition of the fluids must be measured at several locations to detect leaks or causes of impurities, and there is a need for diagnostic instrumentation to anticipate malfunctions and reduce maintenance operations. In summary, there are many new tasks for the control system which are presently not required in existing land-based power plants.

Generators. In terms of size and power output the required generator is clearly within the state of the art and requires no major development. However, the generator size and weight are a function of operating speed. The generator efficiency is assumed to be on the order of 98 percent, and it does require a cooling system. Because of the possibility of corrosive fluids, the generator may have to be cooled by a sealed system, which would be somewhat more expensive than for normal units of this size.

Pumping System

Pumps. For the OTEC power plants, three pumps are required: a cold-water pump, a hot-water pump, and a working-fluid pressurization pump. The working-fluid pump resembles a boiler feed pump. It operates at a moderate temperature of about 50°F and has a relatively low pressure increase and a fairly large quantity of fluid. This pump is entirely within the present state of the art. On the other hand, the hot- and cold-water pumps must deliver very large quantities of water against a relatively small head. As a result, the specific speed is very high and their shaft speed is unusually low. Due to the modular construction of the power plants, the capacity for each pump is reduced, but it is still very high—the needed pump dimensions and the quantity to be delivered exceed the largest pumps ever built.

There should be a model test of the entire pumping system to ensure that the assumed pump efficiency and the system's resistance can be achieved. The pipe dimensions, heat exchanger, flow inlet, and discharge are all unusual and very large. The UMass system requires larger quantities of warm-water flow because of the heat exchanger recommended, and the SSPI system requires an equivalently large cold-water flow because of its proposed configuration. The size of the pumps will require a substantial time for delivery and perhaps special provisions for shipment or a manufacturing plant located with access to the open ocean.

Pump Power Plant. The cold- and warm-water pumps operate at relatively slow speed, on the order of 45 to 60 rpm. Special attention must be paid to the bearings, particularly the thrust bearing, because of the size of the pump and the driving system. Both electric and turbine drives have been suggested. In either case, a reduction gear may be needed to reduce the size and cost of the prime mover. The reduction gear will be both large and costly, and a slow shaft speed will be needed on the pump side. Thus, the number and location of pumps for each of the modules must be carefully analyzed.

Pump Controls. The controls for the pumps require particular attention because the pumps must be operated before the entire OTEC power plant can deliver power. As a result, a control system must be developed to provide for start-up power. In addition, the pumps are connected to a large piping system. The entire system must be analyzed for fluid-flow resonances to avoid pressure pulsations and associated feedback. Special provisions may need to be incorporated in the pump control system. The time required for start-up and shutdown must be carefully analyzed in order to avoid excessive pressures in any part of the system. While such types of analysis are not unusual, the magnitude and size of the flow quantity, the pipes, and the exposure to ocean dynamics make this a new and complex part of the system.

Materials

Choice of Materials. The choice of materials has been discussed in the various reports in substantial detail. The main problems regarding materials in an ocean environment are corrosion and fouling. Fouling is usually caused by micro-organisms or by deposits of various materials on the surface of the structure. There are also inorganic materials such as sulfates and carbonates which are dissolved in seawater. Any deposit caused by fouling affects the performance of the heat exchanger as well as the frictional resistance of the pipes; thus the entire performance of the OTEC power plant can be affected. Seawater itself is corrosive to many materials. The working fluids also can be corrosive depending on which fluid is used. Ammonia is extremely sensitive to water, and it becomes corrosive if small amounts of water leak into the system. The same is true, to a lesser degree, of Freon.

The various investigators have proposed a wide range of materials. For the warm-water ducts, CMU has proposed concrete and SSPI has proposed steel; but other materials like aluminum have also been suggested. There exist coatings for both corrosion resistance and fouling. However, these coatings are presently effective for a limited time only, usually 2 to 4 years, depending on environmental conditions and fluid-flow velocities.

Availability and Performance. A great variety of materials have been proposed. They include steel, aluminum, and concrete for both piping and the major part of the structure. The majority of the materials will require protection against corrosion, fouling, and other sea-inflicted damage. In the case of concrete, particular care must be exercised in the selection and proportioning of coarse and fine aggregate. The processing of the ingredients and mixing affect the quality of the concrete to a considerable extent. Prestressing can be accomplished in watertight ducts which will be filled with grout after tensioning. However, concrete also needs protection against corrosion of the reinforcing elements and polymer additives to provide impermeability. Concrete also may require protective coatings on its surfaces to reduce penetration of chlorine ions and oxygen.

Steel and other materials require protection against corrosion and fouling. Steel is often protected .in ocean structures, and considerable experience has been accumulated. On the other hand, large concrete structures are used mainly in piers and harbors, and there is considerably less experience available for using concrete on open ocean platforms.

The main structural materials are readily available. However, the size of an OTEC power plant may require special provisions for manufacturing. Such provisions have been made in constructing the Ekofisk storage tank, and more of this construction will be done no doubt during the next decade. Nevertheless, it does require special skills and professional supervision of the workforce.

Costs. Cost plays a major role in both design and material selection for an OTEC power plant. Present studies have given considerable attention to heat-exchanger costs since they form the largest single component. It must be emphasized, however, that the cost of construction of the main platform and of the large ducts will require special attention, particularly if the manufacturing is done at a special site rather than in an established manufacturing facility. The selected material and its cost will also directly affect how long the OTEC power plant can be at its site before maintenance and overhaul are necessary. There is a tradeoff between initial cost for materials, including protective coatings against the various actions of the sea, and the later cost for maintaining the OTEC power plant in operation on station. This is particularly critical since fouling or corrosion can increase the friction in the ducts and reduce heat transfer to the extent that power output will be substantially reduced. Details and expected values of the various costs are summarized in Chapters 2 and 6.

Corrosion

Environmental Problems. Protection against corrosion in metallic seawater systems has been established for a long time. Frequently corrosion is caused by electric coupling of dissimilar metals. Therefore, dissimilar materials are to be avoided wherever possible. There are corrosion-resistant materials—for instance, titanium—but their price is usually so high that they can be used only in special applications. Carbon steel is widely used, but it requires protection. This protection consists of various coatings, a cathodic system, or a combination of both. Cathodic protection is an electrical method of preventing galvanic corrosion; it is used widely for the protection of submerged components. Such protection would be desirable near seawater pumps, screens, and similar components. The coatings can reduce the amount of current required for cathodic protection. Both the coatings and cathodic protection systems require inspection at predetermined intervals to ensure satisfactory operation.

Standards. The various material suppliers provide data on the effects of corrosion in terms of maximum depth of pits (in inches or thousandths of an inch) as well as metal loss from corrosion. In order to reduce corrosion, manufacturing processes have been suggested that use clad processes, which employ as the basic material a lower-cost product to be protected by a thin layer of the clad material. One suggestion was to use titanium tubing with tube sheets of low-carbon steel with titanium on the seawater side. The shell would also be made of low-carbon steel in this case. Cladding has also been suggested for the heat exchanger with aluminum alloys. Cladding has been used in many applications–not, however, in the quantities and sizes which have been proposed for the OTEC power plant.

Expectations as to "Clean" Quality of Water. The OTEC power plant requires both a warm and a cold seawater system. In order to provide for reasonably clean water, gates and screens and mechanisms for cleaning the seawater are necessary. This cleaning must be accomplished before the water enters the power module. The intake duct must be protected by a screen to exclude debris sized 1 in. or greater in any dimension. It can be expected that material smaller than this will pass through the heat-exchanger tubes. The proposed screens will be of metallic construction so that they will have a minimum pressure drop. The screens must be cleaned at regular intervals. This cleaning could be accomplished either by divers or by high-pressure water jets. The screen at the bottom of the cold-water intake pipe cannot be cleaned by divers, but would require a submersible unless the cleaning is done automatically by water jet. Fouling of the screen at the cold-water inlet is expected to be minimal due to temperature and pressure at this location. As a material for the screens, stainless steel has been proposed. The screens remove foreign objects but do not affect the corrosion and fouling characteristics of the seawater.

Fouling

Fouling Experience. Fouling is associated with organic mechanisms in seawater. Marine organisms (barnacles) which attach themselves to any surface grow and effectively seal off a small part of the surface from its environment. Fouling, so characteristic of the sea, also leads to a concentration of cells. A favorite site for pits to start is underneath the barnacle and quite often the deepest pits are found there. Fouling is frequently a function of seawater temperature and location. The subject has been treated extensively in the literature. Fouling factors are essentially determined empirically as a result of operating experience, and fouling factors are quoted for different services. In practice, it is sometimes difficult to differentiate between scale, dirt, and biofouling since all reduce heat transfer and frequently occur simultaneously. The general experience has been that fouling increases with seawater temperature and with the relative content of nutrients. It decreases with increasing seawater velocity.

Antifouling Methods. Some metals have a resistance to fouling. They include various copper alloys. The best materials are those containing 90 to 100 percent copper, including the 90-10 copper-nickel alloy. This alloy is also resistant to pitting and crevice corrosion. Some of the copper alloys have their copper content in the range of 70 to 90 percent, such as brass and bronze. They have good fouling resistance but are not comparable to the higher-copper-content alloys. It is noted that large masses of copper put copper ions into the sea.

Recently, a number of antifouling coatings have been developed. Some of these have been very successful. If they are combined with anticorrosion coatings, the coatings must be compatible with one another, and their application requires skilled workers. Usually these coatings are applied in relatively thin layers, and two or more coatings may be necessary.

Mechanical. If fouling cannot be avoided, then cleaning methods must be applied to remove the fouling growth from the surface. Various mechanical devices can be used, particularly in heat exchangers with straight tubing surfaces. In some cases, this requires that the tubes be no smaller than a certain diameter in order to be accessible to the cleaning tool, and it is advantageous to have as many pipes of identical diameter as possible.

Chemical. Fouling can also be avoided by chemical treatment. A number of such systems are commercially available that use chlorine. The treatment consists essentially of providing a chlorine content of 0.5 parts per million (ppm). This value would apply to warm ocean water. Cold ocean water probably could be cleaned with a less concentrated solution. The chlorine treatment is particularly effective in aluminum heat exchangers. If chlorine treatment of 0.5 ppm is selected for both the condenser and the evaporator, then approximately 3000 lb/h of chlorine is used. Such systems have not been used in very large systems or over long periods of time. The long-term environmental effects of such treatments are not known, and they should be explored.

Effects of Velocity. As seawater velocities reach 3 to 6 ft/sec, fouling diminishes. At the same velocities, pitting of the more noble alloys slows down and even ceases. As the velocities continue to increase, the corrosion barrier film is stripped away from the steel and copper-base alloys, while the stainless type and many nickel-base materials remain passive and inert. The complete reversal in the tolerances of metals for the marine environment as velocities change is the source of much seemingly conflicting information on actual experience with marine metals. Velocity effects deserve close study by all designers.

Operation

Staffing. The requirements for the operating workforce are different for the three projects. CMU proposes a fully automatic system whereas the other proposals envisage an operating workforce aboard the platform.

A fully automated plant is not used presently in commercial or naval ship designs. A fully automated plant could be technically feasible, but it would require substantially more instrumentation and automation. It would also require a monitoring facility on land to ensure satisfactory performance and shutdown in case of malfunctions.

If the plant uses workers, there must be provisions for crew quarters and control spaces. Present control systems are designed continually to monitor bearing temperatures, lubrication oil pressures, shaft speed, and various other operating parameters. The controls provide various types of alarms in case of unusual conditions and in some cases automatic shutdown in case of danger.

Also appropriate provisions to facilitate overhaul are necessary. This may include the removal of power modules and the cleaning of various power plant components.

Controls. The requirements for instrumentation and control of the power plant will not involve a particularly complex engineering task or large costs for a staffed power plant. Standard power plant control instrumentation must be provided. However, in one aspect, an additional control system is required. Present nuclear and fossil-fuel power plants regulate the fuel supply as a function of power demand from the electric network system. In an OTEC plant, the energy supply is a function of the temperature difference between hot and cold water, and this may change as a function of many factors. Therefore, the available thermal energy must be matched to the power demand and the capability of the power plant.

The OTEC power plant also requires an external power supply for start-up. The warm-water pump and the cold-water pump of at least one module must be energized and operating until enough vapor is produced to generate power. Since OTEC power plants have a substantial fluid inventory, this start-up will require either a substantial amount of power or a considerable length of time. When internal power is available, the control system must switch from external power to internal power.

Mooring

State of the Art. The three investigators have proposed entirely different concepts for mooring and positioning. UMass and CMU have suggested sea bottom-attached structures while SSPI prefers a floating, self-propelled plant. The motions of the various platforms may vary since some of them resemble spar buoys while others propose large horizontal cylinders. Thus, the motion of the platform in response to waves will be considerably less than that experienced by a ship but will be greater than that of a submersible. The large pipes will assist in dampening horizontal motion. The analysis of motion will require a substantial effort in future studies. It may also involve verification of the proposed analysis in water test facilities and at sea. This technology is not beyond the state of the art because such studies have been made for large buoys. It will, however, require a substantial investment in time and money.

Risks. In order to keep the power plant at its predetermined location, accurate information is needed regarding motion and forces which the power plant structure will impose on the mooring lines. At this time, mooring systems have a rather short life compared to the life expectancy of the power plant. Large buoys have been kept on station for about 1 year, more or less. The mooring system may require regular inspections and replacement of components at certain intervals. A certain amount of risk is involved if the mooring system is expected to provide for long-term, unattended operation without replacement of components. If a dynamic mooring system is used, the requirements for power may increase substantially during high-sea states. At this time, there is no experience for keeping dynamic mooring systems at a predetermined location for a number of years. While a dynamic mooring system is not subject to the malfunction of mooring lines, its pumps and thrusters may lose performance due to fouling, corrosion, or the malfunction of seals. Therefore, such systems also involve a certain amount of risk, simply because they have not been used for positioning large platforms for an extended period of time.

Platform

Design. Concrete has been suggested as a structural material for two of the platforms while the third one recommends steel. Reinforced concrete, aluminum, and steel have all been proposed for manufacture of the pipes. It is quite evident that the investigators have each chosen a different material and a different method of construction. Each of the proposed designs appears to be feasible, and the preferred design may include considerations dictated by the site of the ultimate deployment of the OTEC plant and the place where it is fabricated. It must be stated that the proposed designs go beyond the present state of the art in various areas, and this may require some development to ensure satisfactory performance. The platform and associated systems must have enough reserve buoyancy to reduce risks in case of casualties due to collision or hurricanes.

Manufacturing. Since the designs of the three platforms are fundamentally different, using concrete, steel, and aluminum, respectively, no single method of construction has been recommended. Those systems which involve metal fabrication recommend fabrication in a shipyard, floating the equipment out of docks and subsequently towing it to location. Another method would be to construct components in a shipyard and assemble the entire unit at the site. At present, the construction of offshore oil-drilling rigs provides some similar experience. A concrete offshore structure which rests upon the seabed has been completed. While this experience is valuable, the number of present and contemplated large concrete structures is limited. The CMU design which proposes a reinforced concrete structure resting on the seabed and swiveled at the base in about 1500 ft of water would involve substantial development to ensure proper operation and survival.

Location. The platform design and the manufacturing process are related to the ultimate deployment and location. The location, in turn, may be a function of the supply of warm and cold water and the costs of transmitting power to the user. The present studies do not analyze the transmission and use of power. A decision must be made as to whether the power generated by an OTEC plant will become part of an electric network or will be used on site. The use of these power plants and their location are also a direct function of the minimum temperature difference which can be efficiently converted into electric power. Finally, the location may also affect the proposed maintenance and overhaul system.

Towing. The movement of the power plants from their manufacturing location to the site of use is of substantial interest. At this time, the investigators have not analyzed in detail this part of the system. It is noted that the towing of drilling platforms has had a relatively high accident rate, and this is mentioned here primarily to indicate that the towing of large structures involves many unknown factors. The towing involves stability problems and the necessity for accurate depth-keeping. It also involves a study of survival under various sea and wind conditions. The possibility of on-site assembly would necessitate moving individual components. No doubt this would simplify the towing problem, but it would result in additional expense and technical complexities at the erection on the site.

Power Cable

Power cables are available for the range of power proposed by the various investigators. However, manufacturing facilities are limited at this time, and cables have been used essentially for short distances, on the order of 10 to 20 mi. High-voltage cables are expensive, and when they are used over longer distances, phase corrections must be considered. Direct-current (dc) cables are a possibility for longer distances, but their use as well as the use of dc transmission lines is limited to special applications. Presently, the maximum length of a dc cable is 80 mi. Large cables not only are expensive to manufacture, but also require substantial cost in cable laying. Also they require underwater cable inspection.

Systems Engineering

Critical Elements

From the available technical reports, the critical elements of OTEC plants can be listed in decreasing order of importance as follows:

1. Heat exchangers
2. Cold- and warm-water pumps
3. Cold-water pipe
4. Hull and related structures
5. Mooring arrangements
6. Turbomachinery

From a technical point of view, the critical elements can be listed as follows.

Overall Heat-Transfer Coefficient. Present heat-transfer practice would result in extremely large heat exchangers and consequently high costs. Improvement in the heat-transfer coefficient is thus desirable to total performance. Heat transfer is also affected by fouling factors and, therefore, requires methods for the control of fouling. Heat-exchanger cost reductions are most important to achieve a competitive cost of power generation.

Control of Biofouling. Fouling and control of fouling are not fully understood at this time for all geographic areas under consideration. Fertility of tropic ocean water is generally low, but only *in situ* testing will evaluate the intensity of the biofouling problem.

Corrosion. Corrosion is fairly well understood for present ocean structures. However, none of these ocean structures will be used at its site for the duration of time which is planned for the OTEC plants. Experience with reinforced concrete structures of the physical size contemplated is very limited at this time.

Parasitic Power

The ocean thermal power plant requires a relatively large amount of power for the operation of the auxiliary systems, particularly the pumps. The ratio of gross power to net power is shown in Table 1-1. This ratio is higher for Lockheed's baseline design than the same ratio quoted by the three other investigators. The reason is that Lockheed took a more conservative approach in assessing power requirements for all the auxiliaries. There is a tradeoff between cost and auxiliary requirements. The water pumps are the main consumers of power. If their specific speed is reduced, their efficiency increases slightly; but their shaft speed reduces and their impeller size increases, so this leads to an increase in cost.

There is also the question of how the efficiency of the piping and pumping system can be maintained over a long time. Fouling would be the main problem which might degrade the efficiency. Fouling could be reduced by increasing water speed, but this would require more power and, therefore it may not be acceptable.

The maintenance of efficient piping and pumping systems will clearly affect the entire parasitic power requirement. Without model test data, it may be difficult to assess the correct amount of parasitic power.

Performance Effectiveness

The performance effectiveness is supported by a development schedule which involves the following phases:

1. Advanced research and systems analysis
2. Systems definition and subsystem (component) experiments
3. Final design, construction, and operation of experimental research facilities

Since possible development schedules extend over extended periods, approximately 15 years, it will be necessary to guess at future technical and material developments to predict performance and performance effectiveness. In the assessment of performance, the availability of a prototype operating in the ocean is a tool needed to provide credibility and to ensure that all factors have been properly considered in performance prediction.

The existing reports indicate that power generation can be achieved with the assumptions made by the investigators. After a prototype has been operated, the existing assumptions and statements need to be reviewed in order to predict realistically the performance of a large OTEC plant.

Cost Effectiveness

The total cost of the OTEC plant depends largely on the size and cost of the heat exchangers. Their size and cost, in turn, are directly affected by the overall heat-transfer coefficient and the available temperature difference between warm and cold water. Therefore, no direct cost comparison can be made unless the plants use the same temperature difference and comparable heat-exchanger data. The Lockheed and TRW reports have analyzed the total cost of the three concepts, and their estimates are more conservative than those of the original investigators. They also have brought the costs of the three systems on a uniform basis regarding the date of the estimate.

It is believed that the cost estimate for such a power plant will be much more reliable after a prototype plant has been operational. At this time, it is very difficult to comment on the cost data since many of the proposed subsystems exceed the state of the art in size and technology.

Technical Requirements by Regulatory Agencies

For an ocean-located power plant, United States Coast Guard (USCG) and American Bureau of Shipping (ABS) requirements should be met. USCG requirements in regard to the staffing and safety of such a power plant need to be established. New ABS requirements may have to be determined for concrete structures. The three principal investigators have not addressed themselves in detail to compliance with regulatory requirements. Lockheed and TRW have mentioned regulatory requirements and have established a base for their view.

In addition to the USCG and ABS regulations, there will be many additional regulatory agency recommendations which must be met. They will relate to fire hazard, contamination, life support, and associated systems. It may be impossible to anticipate all these requirements until an actual application for operation is filed. Recent experiences for LNG (liquefied natural gas) ships and offshore power plants seem to indicate that this may not be a large technical problem, but it will require time, substantial attention to detail, and documentation for the processing of the needed permits.

The requirements for environmental protection are unclear at this time. However, if chlorine is used to keep the pipes and heat exchangers clean and to avoid fouling, some regulation requiring documentation for chlorine control and chlorine content in solution can be expected. The same may be required for protection against large leaks of the system fluid.

Requirements of regulatory agencies are also discussed in Chapters 6, 7, 9, and 10.

Technical Requirements for Inspection

Technical requirements for inspection are closely associated with regulatory requirements, in this case ABS and USCG, for ocean structures. There is a need for machinery and component inspection as well as instrument calibration. The major unknowns at this time are the inspection requirements to assess fouling. These plants will require more extensive inspection than existing power plants because of the type of fluids flowing through the heat exchangers and turbines.

Technical Requirements for Maintenance

The three principal investigators have not addressed themselves in detail regarding the maintenance concept for the entire power plant. All three investigators propose a modular concept and indicate the capability of replacing modules. However, implementation requires different facilities and concepts. It may require the use of submersible equipment or variable ballasting to surface the entire station. Some of the heat exchangers have extremely thin walls in order to get a high heat-transfer coefficient, and these thin walls in turn may require extensive maintenance or replacement.

Technical Requirements for Overhaul

The present investigators have not studied overhaul in detail or considered a dry-docking cycle for major components. Periodic dry docking, regulatory agency surveys, and underwater maintenance in dock are done at predetermined intervals for steel ships. For the proposed OTEC power plant, such overhaul could involve great expense and long periods of outage. Possibly systems could be designed, particularly a concrete hull, which could try to eliminate the necessity for such work; the durability of concrete offers such a possibility.

It is also possible that some of the designs can be repaired and surveyed while afloat. The UMass design can be ballasted to surface, thus permitting maintenance of the evaporators. The design of the power plants must provide for cleaning accumulated fouling from the exterior surfaces of the structure.

At this time, it is not possible to evaluate the capability of achieving the needed overhaul requirement with cost-effective designs. This is an area where major problems can develop because of the large size of the power plants and their unusual configuration. Overhaul procedures and costs will be affected by sea and weather conditions.

Consideration must be given to the probability of other vessels colliding with the OTEC power plant. This may require special overhaul and repair analysis to ensure safety and return of the power plant to full operation. Such possibilities should be incorporated into the overhaul procedures and facilities.

Reliability

Reliability is of prime importance for an OTEC power plant scheduled to be on station for many years. There is usually a tradeoff between achievable reliability and initial cost. For instance, the use of aluminum in heat exchangers makes them subject to corrosion and fouling. Thus, a lower initial investment will result in higher maintenance, higher operating cost, and reduced reliability. Concrete is also subject to fouling, and it will affect the buoyancy of the power plant station and thus could indirectly affect reliability. The stability of some of the structures under varying sea conditions over various lengths of time may become a reliability factor. Methods of mooring are known to affect total reliability. Corrosion and biofouling probably affect reliability of the entire plant the most.

Reliability is also affected by the complexity of the entire plant. Careful design, simple systems, and a minimum number of components are means to improve reliability factors. Such studies should be undertaken for the OTEC power plant. In regard to the cold- and warm-water pumps, reliability can be increased by having at least some redundancy to avoid shutdown.

Summary

The present studies indicate that ocean thermal energy conversion is feasible. There is a need for a large development program to improve technology and reduce cost. Beginning operation of a full-sized power plant may be in the 1990-2000 time frame. A smaller experimental unit may be in operation at an earlier date. Once a smaller experimental unit has operated, more reliable data on performance and costs will be available. Overall judgment regarding the OTEC power plant must consider the technical advances that are likely to occur during the next 20 years in the most critical technologies, and these advances are contingent upon the amount of resources, workforce, and capital available to the program. A technical assessment should also include potential performance increases in competing power plant systems and the associated fuel costs of such systems. A final assessment should include such factors as safety, environmental impact, national economy, and political conditions.

Bibliography

Dugger, Gordon L., editor, APL/JHU, SR 75-2, *Proceedings, Third Workshop on Ocean Thermal Energy Conversion (OTEC)* (August 1975).

Hallberg, C.R., "International Regulatory Authority concerning Ocean Thermal Energy Conversion Devices," prepared for Energy Research and Development Administration under Grant to American Society of International Law (1975).

Higgins, James C., Offshore Power Systems, Jacksonville, Florida, "Ocean Thermal Energy Conversion Plants: Federal and State Regulatory Aspects," prepared for Energy Research and Development Administration under Grant to American Society of International Law (1975).

Lockheed Missiles and Space Company, Inc., Sunnyvale, Calif., Ocean Systems, "Ocean Thermal Energy Conversion" (April 1975).

Lockheed Missiles and Space Company, Inc., Sunnyvale, Calif., Ocean Systems, LMSC-DO56566, *Ocean Thermal Energy Conversion (OTEC) Power Plant Technical and Economic Feasibility*, vols. 1 and 2 (April 12, 1975).

Lockheed Missiles and Space Company, Inc., Sunnyvale, Calif., Ocean Systems, Quarterly Progress Report, "Ocean Thermal Energy Conversion" (January 20, 1975).

Stern, Carlos D., University of Connecticut, "Economic Issues Related to Ocean Thermal Energy Conversion Plants" (1975).

Stein, Robert E., International Institute for Environment and Development, "Ocean Thermal Energy Conversion: International Environmental Aspects" (1975).

Stoel, T.B., Natural Resources Defense Council, "Ocean Thermal Energy Conversion: Domestic Environmental Aspects" (1975).

TRW Systems Group, Final Report, *Ocean Thermal Energy Conversion*, vols. 1 through 5, Redondo Beach, Calif. (June 1975).

TRW Systems Group, Quarterly Report, "Ocean Thermal Energy Conversion Research on an Engineering Evaluation and Test Program," Redondo Beach, Calif. (December 13, 1974).

2 An Economic Assessment of Ocean Thermal Energy Conversion

Carlos D. Stern

Introduction

Ocean thermal energy conversion (OTEC) is a proposed system for harnessing the thermal energy of the ocean that exists in some areas of the world as a result of the difference between warm surface waters and cold deep waters. The temperature differential can approach 40°F. In some tropical zones the surface of the ocean can reach 80°F while at depths of about 4000 ft and below the water is at its most dense, about 40°F. In the Gulf Stream, near Miami, Florida, in February the surface temperature is 75°F while at a depth of 3000 ft the temperature is 44.7°F, a difference of 30.3°F.[1] The proposed OTEC technology would convert the energy available in this thermal gradient to mechanical energy, and from the turning of a turbine electricity would be generated. If successful, OTEC would be a novel way of generating electricity.[2] Since it is the sun that warms the surface of the ocean, OTEC is in essence a solar energy converter.

Any system for using solar energy is inherently desirable because the source of energy is vast in comparison with today's requirements; in addition it is nondepletable, nonpolluting, and free. But solar converters generally have one major disadvantage: the energy of the sun is not available when and as needed. Daily and seasonal fluctuations make it necessary to overdesign the collection system to take maximum advantage whenever the natural energy is available. This in turn requires an investment in storage so that energy can be drawn at night and on overcast days. OTEC would have none of these liabilities, since the ocean itself would serve as a cost-free solar collector and thermal storage medium. For the most part, the energy would be available independent of time of day, of season, and of any other natural variation. There are some temperature fluctuations depending on the season. For example, in the Gulf Stream off Florida the summertime temperature difference is 40°F; but in winter this falls to 30°F. Also major tropical storms can have a rather significant, albeit temporary, effect on surface temperature.[3]

The author wishes to acknowledge the assistance given to him by the members of the Panel on Ocean Thermal Energy Conversion of the ASIL. Particular thanks are extended to Drs. Sheets and Rohsenow for technical guidance on the engineering aspects of OTEC; also to Fred Naef of Lockheed; Ralph Eldridge of the Mitre Corporation; Dr. Winer and Dr. Nichol of A.D. Little, Inc.; Mr. Torben Aabo of Power Technologies, Inc.; and Dr. Norbert Greene of the Metallurgy Department and Dr. Michael Howard of Chemical Engineering Department, University of Connecticut. Any errors and omissions are the sole responsibility of the author.

Even though the costs of operating the OTEC may be very low, since there is no charge for the energy taken from the ocean, there is a significant expense associated with the machinery required to harness and convert the naturally occurring thermal differential. Thus solar energy is not really "free." The challenge is to develop an economic means for extracting useful work from what is a very small temperature differential. Modern electric generating units utilize differentials of 600 to 1000°F. Working with a temperature difference as modest as 40°F entails entirely new equipment of a design and scale not heretofore developed. However, if engineers can find an economical method for extracting useful work from small temperature differentials, many other free sources of low-temperature heat, such as condenser discharges from existing nuclear and fossil-fuel power plants, could be used beneficially.

If OTEC can be commercially developed, it will contribute significantly to the world's available energy. There are many uncertainties, however, both technical and economic, that will have to be resolved before a final assessment can be made. The technical uncertainties include its performance, efficiency, reliability over time, cost of upkeep, and method of delivering the energy to users. The economic uncertainties include the potential for cost reduction if OTEC devices are commercially produced, the overall level of demand for energy at the time OTEC devices are ready for deployment, and the cost of competing energy forms in the future.

As an electric generator, an OTEC will be used in one of two ways. First it can provide power to an existing electric grid, and via electric utility systems to the ultimate consumer. Or, alternatively, it can be linked with a floating electric-using industry as a joint venture, supplying processed raw materials. In the first instance, the market for OTEC is tied to the market for electricity. Therefore, the economic analysis will consider level of demand for power in the future, the cost of meeting that demand in the absence of OTEC, and the institutional arrangements (ownership, financing, subsidies, and penalties) that might affect the cost-benefit analysis of OTEC. In the second instance, the market for OTEC will be tied to the market for the particular processed raw materials to be manufactured, such as ammonia for fertilizer production, aluminum, enriched uranium, and pure hydrogen.

OTEC as Part of the Electric Utility System

Potential Market for OTEC

For at least the past 35 years the demand for electricity has grown at about a 7 1/2 percent annual rate, effectively doubling every 10 years. The combination of sharp price increases following the 1973 Arab oil embargo and the economic recession has caused a slight absolute decline in power demand for the first time since the Great Depression.[4] Even before 1973, some economists had argued that the rapid rate of growth would have to slow down, in part because of the saturation of certain uses and in part because of the substantial negative impact of energy production on the environment. The question now is whether the recent downturn in electricity demand is to be seen as the beginning of a new trend toward modest (1 to 3 percent) annual increases or a short-term aberration from the long-term (7 to 8 percent) trend. Until the situation is clarified, utilities will shy away from fixed long-term commitments such as nuclear plants.

In 1973, the United States had a total installed capacity of some 390 million kW which generated 1.88 trillion kWh. This is about a 69 percent annual load factor. Thus, there exists a substantial replacement market for OTEC plants, even in the absence of any growth. The most recent figures indicate that nationally a 1 to 1 1/2 percent growth may have resumed, but with significant regional variations. The Federal Power Commission is still forecasting an annual growth rate through 1985 of about 6 1/2 to 7 percent. It is quite likely, however, that some long-term trends have been set in motion that will counter resumption of the historic growth rate. Throughout the nation, not only are electric rates rising at an unprecedented pace, but rate structures that traditionally have favored large users by granting quantity discounts are being reversed to penalize large users and to reward the frugal. The national energy policy directed toward "low-cost abundant energy" has been abandoned and replaced by explicit policies to encourage conservation. A concerted effort is being made to develop energy conservation technology, some of which (the fuel cell and total energy systems, for example) is a move away from the large, integrated electric grid with its remote generating stations and long transmission lines.

To sum up, there is evidence in both directions. Power demands may well resume the historic growth trend; or a watershed may have been passed in 1973, sharply restricting the need for new generating capacity and thus limiting the potential market for OTEC.

The Place of OTEC in a Utility System

If OTEC were to supply electricity to the grid in the United States, economics would require that the cost of generating and delivering its power, including amortization of the debt and with due allowance for any special subsidies and incentives, be less than or equal to the cost of equivalent power that could be obtained from alternative sources. Almost the total cost of an OTEC plant will be the amortization of the debt; the fuel costs nothing, and the annual maintenance is expected to be low. Since the more the plant operates, the cheaper each kilowatt-hour will be, there will be a strong economic incentive to operate OTEC plants at their maximum sustainable output.

Figure 2-1 shows the variation in demand for electricity over the course of 1

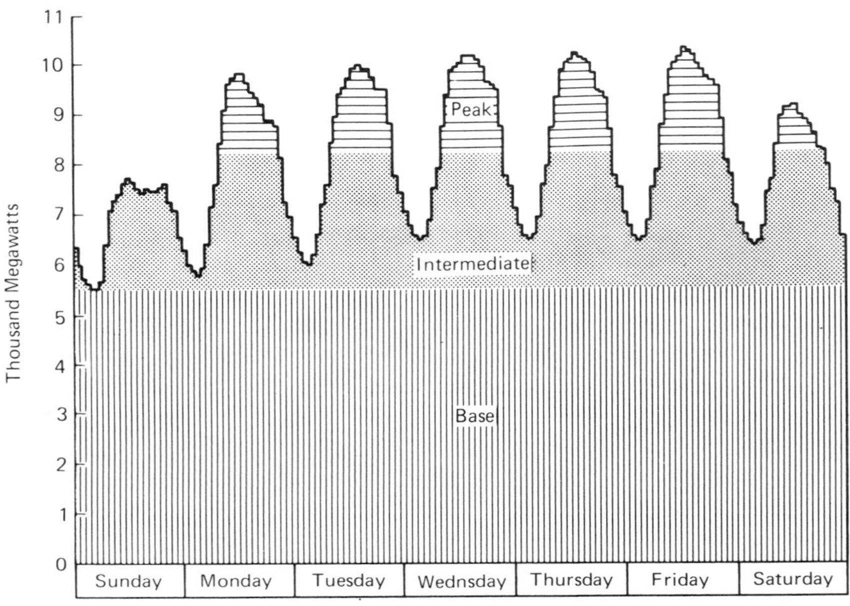

Source: From *The 1970 National Power Survey*, Federal Power Commission, Washington, D.C., 1970.

Figure 2-1. Weekly Load Curve.

week for a typical electric system. As can be seen, the demand for power varies considerably by time of day and day of the week. However, even at the minimum, a certain amount of power is always demanded, and that minimum constitutes the base. The term used by the power industry for plants that operate nearly continuously is *base-loaded*, and the utility will assign its most efficient and most economical generating unit to such near-continuous operation. Generally, plants operating in excess of 7500 hours per year are considered base-loaded. There are 8750 hours in the year. Thus, a base-loaded plant operating at full output of 7500 hours per year would be said to have an "annual plant factor" of 85 percent. It is certain that OTEC plants will be intended to operate at their maximum rated output on a continuous basis, with the only exception being periods of downtime, both scheduled and unscheduled, for necessary maintenance and repairs.

Figure 2-2 shows the integrated electric demand over a period of 1 year. It indicates the relative significance of base-load plants compared with intermediate and peak-load plants in a particular utility system. In this illustration, 28 percent of the total demand (peak) is required 100 percent of the time; and 45 percent of the peak is required 85 percent of the time. This base-load demand is the portion of the load in which the OTEC would fit.

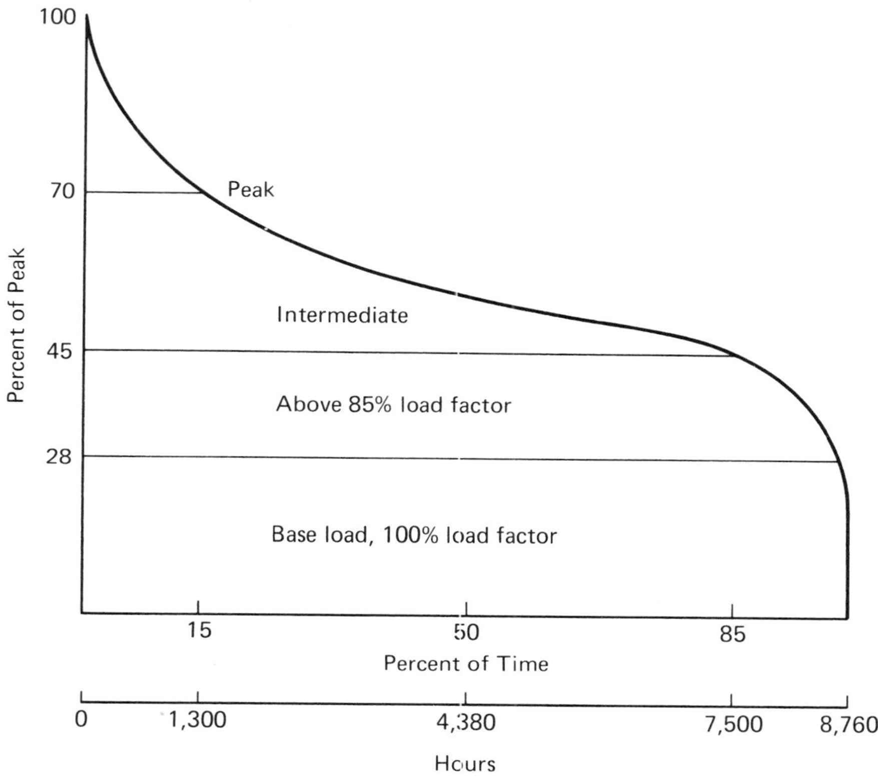

Figure 2-2. Annual Load-Duration Curve.

Since OTEC is base-loaded, its competition in the short term (next 20 years) will be primarily nuclear and efficient coal-fired plants, since these are able to generate the lowest cost per kilowatt-hour. It is extremely unlikely that any new oil or natural gas-fueled base-load plants will be built in this period. Indeed, as part of the national energy policy, the Federal Energy Administration is requiring that even existing base-load oil and gas-fired plants be converted to coal wherever feasible. Therefore, it is inappropriate to compare the cost of OTEC to that of base-load oil-fired plants.

Instead, especially along the East and Gulf coasts, up to and including the early 1990s OTEC's competition will come largely from a backlog of conventional nuclear plants that have been deferred owing to difficulty in raising capital given the uncertainties in the electricity demand picture.[5] As of December 2, 1975, orders for 13 nuclear plants have been canceled and 63 deferred.[6] When these conditions are removed, these plants will be revived before any new equipment is ordered. (When the necessary progress is made to convert coal to a clean fuel, coal will be well placed to increase its share of base-load generating capacity.)

Over the longer term, after the 1990s, OTEC plants will have to compete with other technologies that are capital-intensive and base-loaded too, such as improved nuclear plants (gas-cooled reactors, molten-salt reactors, etc.), fast-breeder reactors, geothermal and land-based solar thermal and photovoltaic plants, and eventually thermonuclear fusion plants.

Relative Cost of OTEC

In assessing the relative cost of OTEC it is important to distinguish between the basic *cost* of producing power at an OTEC (i.e., the total value of labor and materials used in generating and supplying the electricity) and the ultimate *price* charged to the user, including any taxes, tax allowances, or subsidies. The same distinction must be made in assessing the cost of other kinds of power generation.

There are a number of implicit subsidies already in existence such as natural gas regulation by the Federal Power Commission, where the price has been regulated far below the level it would have been under a free market system, and the Price-Anderson Act, which limits the liability of the private owner in the event of catastrophic nuclear plant accident and provides in effect free insurance for power companies owning and operating nuclear plants. Another implicit subsidy is in the area of ownership. Government-owned (municipal, federal, or other) enterprises permit lenders to pay no federal income tax on the interest received. This has the effect of lowering the interest rate on money borrowed by public utilities. Explicit subsidies include the price charged by the government for uranium enrichment, depletion allowances on fuels, levies on imported fuels, and the U.S. Maritime Administration of the Department of Commerce subsidy to U.S. shipbuilders.

Each of these arrangements can have a profound effect on cost comparisons. Moreover, these policies can always be changed. It is quite possible that Congress will enact taxes on depletable fuels as a conservation measure some time in the future, and at the same time it might provide subsidies for natural energy converters. Obviously, the intent of such legislation, when and if it is enacted, would be to bias the choice toward systems like OTEC. The issue before us at this time, however, is how competitive OTEC is now, as measured in normal economic terms.

A basic political or "value" issue is whether the free market is able to allocate resources efficiently over the long term. Of particular concern to some is whether the pace of resource exhaustion may be exceeding the rate of discovery of new reserves and the development of substitutes. Those subscribing to that concern would argue for economic penalties on the depletable fuels and would favor economic incentives for natural energy systems. But natural energy converters are capital-intensive, and capital itself must be seen as a limited resource. A high interest rate is the free market mechanism for indicating the relative scarcity of capital, and at the same time the interest rate serves to allocate resources over different time periods. Given that *time* is also a nonrenewable resource, high interest rates are a way of biasing choices away from capital-intensive investments to less capital-intensive projects. An artificially low interest rate will serve to distort the mix of long-term and short-term commitments.

OTEC, like hydroelectric projects, will be highly capital-intensive to the extent that 90 percent of the costs will be fixed and are thus independent of the actual operation of the plant. The fact that a large investment will be required in advance does not in itself mean that there must be a government subsidy. Investor-owned utilities are quite capable of financing hydroelectric and nuclear projects, as long as the investment falls within what is considered a normal business risk. Obviously, the riskier the project, the more reluctant investors will be to finance it. In this respect, a long payout period can increase the risk to investors.

A high proportion of fixed costs has the economic advantage of eliminating the risk of future production cost increases. The ability to fix future costs is, of course, not unique to capital-intensive plants such as OTEC. It is quite common practice for coal-burning power plants to be committed along with the necessary coal reserves for the 35-year life of the plant. Utilities quite commonly purchase coal mines outright and occasionally the associated transportation coal-haul facilities as well. Where a purchase is not made, a 35-year contract may be negotiated. Thus, for coal plants, the cost of fuel is generally fixed for the life of the plant.[7]

The economic disadvantage of a long payout period is that costs cannot be reduced by curtailing operations. The owner is committed for the life of the project regardless of the actual demand for power, the performance of the OTEC over time, and the arrival of less expensive energy sources. The economic life of an OTEC platform is expected to be 100 years, the mooring 50 years, the power modules 50 years, and the rotating machinery 20 to 25 years.[8] Over this long duration in such an active area of technology, the risk is twofold: (1) new technologies may render OTEC obsolete; (2) the plant components may not survive for the full period.

At the present time, the annual cost of ownership for government-owned utilities, such as REA, municipals, TVA, and Bonneville, is about 9.6 percent per annum, made up as follows: interest payment, 7.0 percent; amortization, 1.6 percent; administrative expenses, 1.0 percent; taxes, 0.0 percent. For private or investor-owned utilities, the annual rate is 15.9 percent, made up as follows: interest payment, 11.0 percent; amortization, 0.9 percent; administrative expenses, 1.0 percent; taxes, 3.0 percent.[9] TRW estimated that their baseline design plant would cost approximately $2100 per installed kilowatt.[10] This is the equivalent of $244.8 million per annum for government ownership, or $377.3 million if the OTEC were to be privately financed. About 80 percent of the electricity generated in the United States today is at privately financed utility systems.

In the example just given, 3 percent per annum would be saved if the OTEC were government-financed because government projects do not pay taxes. On the other hand, the 3 percent taxes foregone are a direct loss to the federal Treasury. The 3 percent thus constitutes a transfer from the tax-paying public at large to the beneficiaries of OTEC, in this case not the nation but the power users in the area to be served by the utility. The true costs to society of the OTEC—that is, the resources committed to the project—are not dependent in any way on the mode of financing. Yet the net cost in dollars can vary by 50 percent depending on whether the project is publicly or privately financed.

Cost Estimate for OTEC

Cost estimates for equipment that has never been constructed and operated must be considered tentative. Therefore, it is not surprising that the cost estimates provided by research studies have varied widely, ranging from as low as $200 per kilowatt and $300 to $750 per kilowatt to $2100 per kilowatt (TRW baseline design) and $2600 per kilowatt (Lockheed baseline design).[11] An independent estimate by Alvin Strauss of the University of Cincinnati calculated $3000 to $3900 per kilowatt.[12] When these investment costs are translated into bus bar electric costs, the range is from 4 mills per kWh (Anderson) and 4.6 to 7.6 mills per kWh (Heronemus), to 35 to 51.8 mills per kWh for government-owned and investor-owned utilities respectively (TRW baseline), assuming a 90 percent plant factor, and 34.0 and 42.8 mills per kWh for a 90 percent plant factor (Lockheed) (for an 80 percent annual plant factor, this increases to 39.4 and 58.3 mills per kWh respectively). Alvin Strauss' estimate is between 89 and 112 mills per kWh depending on the specific design. All these costs are based on 1974 dollars.

To put these costs into perspective, nuclear power is about $500 to $700 per kilowatt, or about 18 to 23 mills per kWh. Coal-burning plants range from $400 to $500 per kilowatt, or 20 to 30 mills per kWh.[13] These estimates are all for privately financed utilities, and this would be compared to the 35 to 51.8 mills per kWh estimate for OTEC to which must still be added the cost of transmission by undersea cable. According to the Lockheed and TRW data, their baseline OTEC plants would be more expensive than the competition today.

The gap could be narrowed in one of three ways (or a combination of these): (1) cost reduction in OTEC technology as experience is gained with early prototypes; (2) government financing and/or substantial subsidies for OTEC; or

(3) continuing cost escalations in nuclear and fossil-fuel generating plants and fuels.

(1) It must be emphasized that the Lockheed and TRW designs are of a baseline nature and intended for early implementation. The proponents' expectation is that engineering improvements and manufacturing innovations will lead to substantial cost reductions. Lockheed, for instance, estimates that the first OTEC will cost about $3700 per kilowatt; their baseline multiunit production cost cited earlier was $2600 per kilowatt. They believe that with design improvements and mass production, capital costs can be reduced to between $1375 and $2020 per installed kilowatt, giving an energy cost of between 17.7 and 25.5 mills per kWh.[14] The proponents contend that as experience is gained, production costs must come down. They feel that any comparison between OTEC and nuclear and coal generating plants should make proper allowance for the fact that the one is an infant and the other a mature technology. On the other hand, experience does not inevitably yield economies. Nuclear costs rose sharply as experience was gained with the earlier units. The first large nuclear plants, Oyster Creek and Browns Ferry, were contracted for at $140 and $110 per kilowatt respectively. In the 8 years since, nuclear costs have climbed to $500 to $700 per kilowatt.

(2) The argument is made that because OTEC devices are capital-intensive, the government will have to subsidize or to finance outright their construction. The government has subsidized the development of nuclear power, and there have also been subsidies in fossil fuels (e.g., the depletion allowance). Therefore, it is virtually certain that the government will also be called upon to subsidize the initial development of this new technology. But, at the commercial deployment stage, OTEC may have to compete for private funding as nuclear and coal do today. If it fails to be so funded and if the federal government owns and operates it, however, some restructuring of the private utility sector will result. The federal government would be moving into the business of power wholesaling, with the private utilities serving as distributors and retailers.

There are certainly good reasons to prefer OTEC to systems that consume nonrenewable resources, and this can be accomplished, as we have seen, either by direct subsidy to OTEC and other natural energy systems or by a depletion tax on nonrenewable fuels.

(3) Prior to the Arab oil embargo and the tremendous increase in energy costs, there was little interest in solar energy for the obvious reason that crude oil at $2 per barrel, natural gas at $.25 per 1000 ft^3, and coal at $.20 per million Btu were cheap enough to deter competition. These costs have risen fivefold since 1973 and are likely to continue to increase. At the same time, the capital costs of the generating plants have also risen substantially in part because of the extraordinary inflation in the industry and in part because of the added environmental controls. In the same period, all the new energy sources—oil shale, tar sands, the liquid metal fast-breeder reactor, coal gasification, to mention just

a few—have also experienced cost increases of several hundred percent with no end in sight. As these energy sources continue to increase in cost, OTEC will become more competitive, provided, of course, that it is better able to resist the same inflationary trends.

Areas of Greatest Cost Uncertainties

The projected cost of OTEC is subject to major uncertainties, any or all of which could have an important impact on the total cost of delivered electricity. These are: (1) the heat exchangers, pumps, and associated equipment; (2) the effect of biofouling on the overall output; (3) the cold-water intake structure; (4) mooring systems; (5) undersea transmission to shore; and (6) operating reliability.

(1) The heat exchangers are by far the largest single cost component of an OTEC, accounting for about 50 to 60 percent of the total cost of the completed power plant.[15] Special assumptions have had to be made in the preliminary studies regarding the type of materials, the design, and the manufacturing process. Both TRW and Lockheed have based their analysis on the use of titanium, the material most ideally suited for their purpose. However, both studies seem to have overlooked the fact that deployment of OTEC on a commercial scale would have such a profound effect on the demand for this scarce metal that a major price inflation would almost certainly result.

The baseline Lockheed OTEC design for a 160-MW plant would call for 17.6 million lb of titanium. The total U.S. titanium tube production in 1974 was but 2.9 million lb, and the total production of titanium ingot was 72.3 million lb. The capacity of the industry was 80 million lb.[16] According to this configuration, a 1000-MW OTEC would require 150 percent of the total U.S. 1974 output of titanium, or 40 times today's total titanium tube production.

It may not be realistic to base the cost estimates for commercial OTEC power plants on the assumption that mass production can be undertaken while ignoring the likely impact on the price of titanium.[17] In fact, Lockheed did the opposite. According to the Lockheed study, baseline cost estimates ". . . took consideration of mill runs and bulk discounts as in heat exchangers . . . [where] . . . quantities . . . should influence a potential cost reduction."[18]

In order to improve the economics, Lockheed estimates that in the future they can reduce heat-exchanger costs while increasing electric output by a combination of less costly fabrication techniques and less costly materials.[19] Normally, cheaper fabrication and cheaper materials do not improve performance. Even without cheaper manufacturing, assembling the heat exchangers will be a difficult challenge. The largest heat exchangers constructed for electric utilities today—the 1200-MW size—approach 30 ft in diameter, whereas the baseline OTEC design of Lockheed requires two, each of 72 ft diameter, for a 40-MW plant.[20] Furthermore, Lockheed projects a cost reduction by going from titanium heat exchangers in the $1600 per kW range to aluminum in the $700 per kW range with some basic changes in design.[21] There is some question as to whether aluminum can stand up reliably to the corrosive forces to which it will be exposed.[22] In addition, if aluminum is used in the heat exchanger, it will be necessary to use aluminum in the power module too, which will add to the cost. Lockheed did not calculate the effect of that cost increase on the cost of the total system.[23] Without question, the development of an economic, highly efficient, reliable, and mass-producible heat exchanger is one of the most vital research and development tasks for OTEC.

A second area for development is to reduce the parasitic power requirements which presently absorb about 20 to 40 percent (33 percent for the Lockheed baseline design) of OTEC's gross electric output.[24] Lockheed states that improvements can be accomplished by "increasing the efficiency . . . with negligible increase in the cost of the hardware."[25] The engineering effort required to achieve this will likely be substantial because the "pump dimensions and quantity [of water] delivered . . . exceed the largest pumps ever built."[26] For example, each 40-MW OTEC plant would need some 1.2 million gal/min of warm water and 1.6 million gal/min of cold water from each of its two pumps.[27]

(2) "Biofouling is one of the unexplored factors with a potential to impair the performance of OTEC plants, rendering them economically non-competitive," according to TRW.[28] The report adds that "massive fouling can be expected in the photic zone."[29] In the tropics, the photic zone can extend to 600 ft below the surface. Thus, most of the power plant would be exposed to sunlight. In any case, marine growths do occur well below the photic zone.[30] Since tropical waters are not very active biologically, one of the benefits hoped for from OTEC is that a recirculation of nutrients from the deep-water zones would enhance biological productivity. If that were to occur, the rate of biofouling would be increased.

Biofouling can seriously degrade the already low thermodynamic efficiency of OTEC plants in two ways: (1) by increasing friction, which forces the pumps to work harder to move the necessary volume of water; and (2) by degrading the overall heat-transfer coefficient of the heat exchangers.

To see the effect of fouling on the economics,[31] see Table 2-1.

ERDA is funding research to study the characteristics of biofouling in tropical waters. There is also industrial research in progress developing processes to counteract and limit fouling by chemical means, and to remove it by mechanical means.[32] At this time there is no question that biofouling is one of the most serious unresolved issues in the economics of OTEC.

(3) At the low thermal efficiencies at which OTEC must operate, it is highly desirable to utilize the coldest water attainable. The temperature of the water drops with increasing depth such that at a specific location an operating temperature differential of 31.3°F is found at the 1500-ft level, 34.8°F at 2000

Table 2-1
Effects of Fouling

Fouling Factor	*Description*	*Effect on Cost of Heat Exchanger*	*Change from Design Basis*
0.000	Clean as delivered	$162/kW	–11.5%
0.125	Optimistic	170	– 7.0
0.250	Design basis	183	no change
0.500	Conservative	200	+ 9.0
1.000	Pessimistic	230	+26.0

ft, 38.5°F at 3000 ft, and 39.3°F at 4000 ft.[33] Every 2°F loss in available temperature difference entails a 9 percent loss in electric output, the equivalent of a 10 percent increase in kilowatt-hour cost.[34] Thus, there is an economic tradeoff between some sacrifice in output offset by some cost saving in the design of the cold-water pipe.

A 4000-ft intake, almost 1 mi of continuous pipe between 50 and 120 ft in diameter, would pose some very special engineering problems. The Lockheed design assumes a 1500-ft concrete pipe, while TRW examined and rejected concrete because of its weight and the danger of cracks.[35] TRW prefers fiber glass. The pipe is designed to have a 100-year life—essentially to be a permanent structure—because repairs to the intake pipe would be extremely costly, in terms of both downtime for the plant and actual dollars.

(4) "The basic problem [with mooring] is that performance at sea of a very large floating system involves non-linear phenomena which are imperfectly understood and very difficult to model mathematically."[36] Yet in order to transmit electricity to shore, a nearly rigid mooring is essential. The TRW report states: "If a high voltage DC power cable is to extend from the OTEC plant, the mooring must necessarily be nearly rigid, since the ability of a cable to stretch is limited."[37] To illustrate, if the plant is located in waters about 1 mi deep, its mooring cable will have to be nearly 5 mi long to permit a 5:1 catenary. Simple geometry shows that a 5-mi-long mooring cable can allow a significant wander radius, anything but a rigid coupling. Some of the Gulf Stream sites under investigation are, in fact, 20,000 ft, or 3.5 mi, deep. The present state of the art of mooring systems for large buoys provides a life of only about 1 year.[38] Therefore, a significant research effort will be needed to develop a system with a reliable 50-year life, as proposed by Lockheed.

To get around this problem, TRW preferred a dynamic positioning system, utilizing the discharge velocity from the boilers and condensers to maneuver the plant against ocean currents. TRW has not calculated the power consumption needed for station-keeping. They state: "A correlation between the horsepower required and the displacement tonnage of the vessels planned for [OTEC] . . . has not been made." It is unfortunate that this estimate is missing because with the thermal efficiency already very low, any further degradation of net electric output could have serious economic consequences. There is also a technical problem in the sizing of the positioning jets. TRW cautions:

> Horsepower requirements for dynamic positioning must represent a compromise and carry a certain risk. It is obviously impossible to provide enough horsepower for station keeping if certain combinations of wind, waves and current are allowed to build up against the surface vessel.[39]

The hydrodynamic drag can be as great as 15 million lb in the Gulf Stream. Therefore, dynamic mooring probably will not be economically feasible in locations with steady currents on the order of 3 kn. Tropical waters are subject to violent storms resulting in high-sea states. Station-keeping under these conditions is beyond the present state of the art.[40] Even if dynamic positioning were to succeed, the amount of power diverted to keeping the OTEC in position during such storms would be substantial and would cause a corresponding unscheduled loss in the net electric output.

(5) No cost estimates were made of undersea transmission because the technical requirements exceed the current state of the art and because the contract with NSF did not require an analysis of transmission. Present technology for undersea alternating-current (ac) transmission is limited to 20 to 30 mi. Longer transmission distances require direct current (dc) which in turn requires an ac/dc inverter at each end of the line. The two inverters cost about $70 to $80 per kW.[41]

The first long-distance undersea dc power cable is being constructed between Norway and Denmark. The Skagerrak cable will be 80 mi long, consisting of four cables each with 250-mW capacity and a working voltage of 250 kV. A new type of cable had to be designed for this purpose, and the mechanical challenge of laying this cable in water depths up to 2000 ft has to be surmounted.[42] The line has not yet been energized; therefore it will be some time before operating experience has been accumulated. Another dc line (266 kV), about 20 mi in length, between Sweden and Denmark has had several failures because of mechanical interference with ships' anchors and trawlers' nets.[43] The Skagerrak line is estimated to cost about $1 million per mile.[44] Today's undersea power cables are from shore to shore, their terminals are land-based, and all the couplings are designed to be rigid. OTEC will require the one dc terminal to be fixed on the ocean floor at a depth of several thousand feet. A flexible coupling, such as swivel and special cable, will have to be designed to link the undersea terminal to the OTEC. The power coupling to the OTEC also will have to be of a special design not heretofore developed because the power cable must be kept rigid.

These two items are unique to OTEC and will therefore not be designed except as part of an overall OTEC research program. Only experience with prototypes will provide some indication of expected reliability, maintenance, and repair costs. Until these prototypes exist, there is no way to establish with confidence the cost of transmission. In any case, the cost per kilowatt-hour will depend on the deployment of OTEC devices, the number connected to a single undersea transmission cable, the actual distances involved, the undersea geology, the ability to moor the plant as rigidly as possible, and the need for routine maintenance and replacement. The cost of the transmission will then need to be added to the capital cost to produce a total, delivered kilowatt-hour figure.

(6) Since the cost of electricity from OTEC is directly related to the hours of operation, it is essential that the plant be base-loaded and operate at full output as near to 100 percent of the time as possible. Reliability is a question not only of engineering but also of economics, since increased reliability can be had at a price. More frequent routine maintenance and the overdesign or duplication of key components to provide redundancy are ways of enhancing plant reliability. Lockheed's power modules would be returned to a land-based facility every 2 years for major overhaul.[45] Since each module costs $81 million, a spare module would increase the capital cost by some $500 per kilowatt.[46] It might be possible for two or more OTEC devices to share a spare module and reduce the cost accordingly, but this would in turn reduce the reliability.

In the same way, redundancy can be, and usually is, built into the transmission system so that the interruption of a single cable will not result in a complete loss of power. The important point is that the reliability of an entire OTEC system can be only as good as the weakest link, be it the swivel coupling of the transmission line or corrosion or biofouling of the heat exchangers.[47]

It was assumed earlier that the heat source would be both stable and reliable. Yet the very operation of the OTEC could lower the intake temperature through recirculation of the plant's own exhaust stream. It will require some sophisticated computer modeling and actual tests to ensure that this does not take place. It is known that some warm ocean currents such as the Gulf Stream shift their position periodically. If the currents shift, the moored OTEC may find itself no longer situated in the zone of warmest surface waters, with a resultant impairment in electric output.

On-Site Electrochemical Processing

Because of the great technical difficulty in transmitting electricity from plant to shore, it has been proposed that the OTEC plant serve a companion sea-based high energy-using processing plant directly, such as uranium enrichment, aluminum reduction, or hydrogen production. In this mode, OTEC plants lose one of their major attractions because even though electricity cannot be stored, products made from electricity can easily be stored, permitting a less constant power supply such as a land-based solar plant more closely tied to the natural solar cycle. Land-based plants have major advantages when it comes to transporting raw materials, disposing of residuals, environmental and safety controls, and repairs and maintenance.

Uranium Enrichment

The most promising joint OTEC-industrial venture could be uranium enrichment because it would require the least transport of raw materials and the least generation of residuals. On the other hand, there are some major uncertainties. On the safety side, it is unknown whether national policy will permit such a vital service to be conducted at sea, far enough from the shore to make transmission of electricity to the mainland infeasible. Furthermore, if new technology for uranium enrichment—the gas centrifuge, for example—were to become the accepted enrichment process, the electricity used would fall substantially below what it would be for the gas diffusion process (the commonly quoted reduction is of the order of 90 percent). Further, nuclear plants would compete with OTEC. Hence there is an inherent contradiction in anticipating a significant market for OTEC to produce fuel for nuclear plants.

Aluminum Reduction

Electrolytic reduction and refining of metals such as aluminum, titanium, and lithium also present a potential market for OTEC-industrial ventures. A number of OTEC devices would probably need to be linked together to serve one such industrial facility, and a larger platform would have to be deployed to house the potlines and other industrial equipment which, at least in their normal terrestrial configurations, require much space. Metals processing would also entail the movement of large tonnages of raw materials, waste material, and finished products. The problems associated with a project of this scope located in tropical oceans will require far more detailed evaluation before a determination as to its feasibility and its desirability can be reached.

Magnesium production does appear to be a practical option, since the magnesium is recovered directly from the seawater and therefore no supplies of raw materials would be required.

Hydrogen Production

The Institute for Gas Technology is studying the feasibility of the electrolytic dissociation of seawater. Hydrogen is a valuable chemical useful both as a clean fuel and as an important chemical feedstock for products such as ammonia and other compounds.

Unfortunately, hydrogen production at sea would involve two stages of efficiency loss. The first would occur during electrolysis which is perhaps 90 percent efficient. However, 90 percent of that output is oxygen, which would have little commercial value. The net efficiency of hydrogen production would thus be about 9 percent. In terms of the energy value of the hydrogen as a percentage of the energy value of the electricity, the second efficiency loss occurs when the hydrogen gas is cooled to –400°F so that it can be transported to the market as liquid hydrogen.

The efficiency of the refrigeration steps is at best 66 percent, giving a total system efficiency for liquid hydrogen from electricity of 6 percent.[48] Calculations show that 190 MW of OTEC capacity can produce 70 tons/day of liquid hydrogen.[49] On a Btu basis, this conversion confirms the 6 percent overall cycle efficiency estimated above. The already low efficiency of OTEC combined with the low efficiency of liquid hydrogen production–6 percent of 2 percent, or 0.12 percent–strongly suggests that the hydrogen will be very expensive on a Btu basis. Preliminary calculations are that the delivered price of hydrogen would be about $15 per million Btu if the OTEC electricity is at 27 mills per kWh (government ownership) and about $22 per million Btu at 37 mills per kWh (investor ownership).[50] The studies being done by the Institute for Gas Technology will help firm up these cost estimates.

For purposes of comparison, at the present time, the price of natural gas in the producing states is about $1.50 to $2.00 per million Btu; it is estimated that if natural gas were deregulated, its price might increase to about $3 per million Btu. Synthetic methane gas made from coal is predicted to sell in the same $3 price range.

Ammonia Production

Ammonia production is based on hydrogen, but it would not require the liquefaction stage; therefore it would be a more promising industrial partner for OTEC. Ammonia is used, among other things, for fertilizer production, and there will certainly be a growing worldwide demand for ammonia-based fertilizers. The ammonia plant may well be quite compact, enabling it to be located within the OTEC itself with corresponding cost economies.

Conclusion

This chapter has described direct economic factors associated with OTEC. In addition to the primary costs and benefits, there are external effects that must be considered as well. Some of these are beneficial. For example, fishing might be improved as a result of the recirculation of the nutrients (upwelling effects).

OTEC plants could serve as stations for data gathering or for sea rescue. Negative effects include possible damage to marine organisms because of thermal and chemical discharges, entrainment, and impingement. Interference with established shipping lanes might also be a cost of OTEC. Possible long-term effects on the global climate is still another. All these external effects are discussed in more detail in other chapters. The purpose of mentioning them here is to make a case for their being quantified to whatever extent possible and then incorporated in the overall economic analysis of the OTEC technology.

An economic analysis at this early stage in the development of a technology like OTEC is, on the one hand, premature in that there are many uncertainties, as discussed in the preceding pages. On the other hand, it should inform the decisionmaker of the potential competitiveness and the possible place of the new technology in the larger national and international scheme of energy production, as well as the direction that further research should take. Going ahead with any technology toward the prototype stage, given the high costs of research and development, must inevitably be done at the expense of other government priorities including the development of other new energy sources. Therefore, each of these new technologies should be examined and reexamined all along the way as rigorously as possible to ensure that only the most promising continue to be supported.

Notes

1. Alvin Strauss, *Solar Sea Power Plants (SSPP), A Critical Review and Survey*, Goddard Space Flight Center, Greenbelt, Md. (September 1974). Available as NTIS publication no. N 75-11459.

2. Ibid., p. 6. Strauss estimates that OTEC "has the potential for supplying 100% of the U.S. electric energy requirements in the not too distant future."

3. Ibid., p. 22*ff. Also*, personal communication with Dr. Herman E. Sheets.

4. The Federal Power Commission publishes statistics on the U.S. electric power industry on a regular basis in *FPC News.*

5. *See* Chapter 11.

6. Report on the status of orders for nuclear power plants by Edison Electric Institute as quoted in *The Hartford Courant*, December 2, 1975.

7. For example, in a November 26, 1975, report to the shareowners of American Electric Power Company, chairman of the board Donald C. Cook stated: "*Fuel Supply:* We expect to have our own coal-mining capacity of over 27 million annual tons by 1980, about three times this year's production. Toward this end, we own or have access to a reserve of over 3.3 billion tons of recoverable coal. *Coal delivery:* We have taken the necessary steps to assure

delivery of the above coal to our power plants. This program includes the acquisition of 3,200 railroad cars, 240 river barges and 16 towboats and the construction of a major coal transfer terminal on the Ohio River in Illinois."

In recent advertisements Utah Power and Light Co. stated: ". . . the company estimates that it already owns or has proprietary rights to sufficient coal supplies to fuel all steam generating plants planned for construction between now and the year 2,000 . . . through the life of the plants. This helps to assure Utah Power's long term fuel supply for its generating plants–at relatively low cost."

8. "Ocean Thermal Energy Conversion (OTEC) Power Plant Technical and Economic Feasibility." vol. 1, Technical Report, Lockheed Missiles and Space Company Inc., Sunnyvale, Calif. Technical Report NSF RANN (12 April 1975), p. 2-38.

9. TRW systems group, "Ocean Thermal Energy Conversion." Final Report (5 vols.), Redondo Beach, Calif. Technical Report NSF C-958 (June 1975), vol. 1, p. 3-25.

10. Ibid.

11. Cited in Strauss, op. cit., p. 6, p. 11; TRW, op. cit., p. 3-25; Lockheed, op. cit., pp. 2-96 and 2-103.

12. Strauss, op. cit., p. 21.

13. Information given in private communication with Northeast Utilities Corp. *See also* Strauss, op. cit., pp. 7 and 9.

14. Lockheed, op. cit., pp. 3-27 and 2-90.

15. Herman E. Sheets, "Ocean Thermal Energy Conversion Plant," Technical background paper, University of Rhode Island (November 1975), pp. 10 and 13. *Also* Lockheed, op. cit., p. 2-90.

16. Lockheed, op. cit., p. 2-26. Private communication with Titanium Metals Corporation of America. Also *Minerals Yearbook 1973*, vol. 1, p. 1233, U.S. Bureau of Mines, GPO, Washington, D.C., 1975.

17. Titanium is a plentiful metal in the raw state, although suitably concentrated naturally occurring ores are relatively rare. At present the demand for titanium is limited to special applications, primarily in the defense area. The metal is expensive because of its large electric energy requirement. The question is whether a vastly increased demand would increase costs, the usual consequence, or lead to cost reduction as a result of economies of scale from increased output. Titanium production requiring 20 kWh/lb is extremely energy-intensive, even more so than aluminum production. The 17.6 million lbs per 160 MW OTEC would thus require 352 million kWh, 25 percent of the entire first year's electric output of the OTEC.

18. Lockheed, op. cit., pp. 2-91 and 2-92.

19. Ibid., p. 2-104.

20. Ibid., p. 3-9.

21. Ibid., p. 3-12.

22. Sheets, op. cit., p. 37. Personal communication with Combustion Engineering and Dr. Norbert Greene, Metallurgy Department, University of Connecticut.

23. Lockheed, op. cit., p. 3-19.

24. Ibid., pp. 2-49 and 2-104.

25. Ibid., p. 2-104.

26. Sheets, op. cit., p. 18.

27. Lockheed, op. cit., p. 2-37; and TRW, op. cit., p. F-44.

28. TRW, op. cit., vol. 4, p. 2.

29. TRW, op. cit., p. F-13.

30. Personal communication with Dr. William Aron, National Oceanic and Atmospheric Administration, Washington, D.C.; and TRW, op. cit., vol. 5, p. 11.

31. Lockheed makes a 5 percent allowance for biofouling in the condenser and 10 percent on the evaporator in their baseline design. Lockheed, op. cit., pp. 2-26 and 5-15.

32. Sheets, op. cit., pp. 24-26.

33. TRW, op. cit., vol. 1, p. 3-6.

34. Strauss, op. cit., p. 22.

35. TRW, op. cit., vol. 1, p. 3-2; Lockheed, op. cit., p. 2-60.

36. Lockheed, op. cit., p. 4-8.

37. TRW, op. cit., p. F-53.

38. Sheets, op. cit., p. 28.

39. TRW, op. cit., p. F-54.

40. Lockheed, op. cit., p. 2-64 (citing Heronemus); Sheets, op. cit., p. 28.

41. Personal communication from Dr. Nichol of A.D. Little, Inc., Cambridge, Mass.

42. B.R. Nyberg, K. Herstad, and K. Bjørløw-Larsen, "Numerical Methods for Calculation of Electrical Stresses in HVDC Cables with Special Application to the Skagerrak Cable," *IEEE paper* T74-461-0.

43. Personal communication from Mr. Aabo, Power Technologies, Inc., Connecticut.

44. Lockheed, op. cit., p. 4-31; and personal communication from Dr. Winer, A.D. Little Inc., Cambridge, Mass.

45. Lockheed, op. cit., p. 2-4.

46. Ibid., p. 2-94.

47. Sheets, op. cit., pp. 37 and 38.

48. Personal communication from Connecticut Natural Gas Company operators of a liquefied natural gas (LNG) facility indicates that about 25 percent of the gas is consumed in the refrigeration and storage cycles. LNG must be cooled to –259°F, and liquid hydrogen to –400°F.

49. Personal communication from Lockheed.

50. Ibid.

3 International Jurisdictional Issues Involving OTEC Installations

H. Gary Knight

Introduction

The siting of OTEC devices in the ocean presents a number of international legal problems, most of which are jurisdictional in nature, i.e., concerning the rights of nations or individuals to install and operate OTEC devices and the restrictions which may be applied to them in the conduct of such activities. Such problems evolve from the concept of scarcity of resources, with jurisdictional rules developing in response to pressures on the use of ocean space. This chapter will identify and assess the present and evolving legal rules and institutions applicable to OTEC siting, with particular emphasis on the legal status of the resource being exploited and the impact of that exploitation on the law of the sea. The chapter assumes that such installations might be located anywhere in ocean space, including territorial waters, economic resource zones (if and when established), and areas clearly beyond the jurisdictional reach of any coastal state under existing international law.

General Jurisdictional Issues

Impact of Site Location

This section consists of three subparts: (1) a brief exposition of the *existing* international law of the sea relevant to OTEC installations; (2) a discussion of the on-going law-of-the-sea negotiations and the most likely outcomes of those negotiations in terms of *changes* in the legal framework governing the use of ocean space; and (3) an examination of the impact of *unilateral actions* on the developing law of ocean space, again as they may impact OTEC deployment.

Present International Law. The rules applicable to activities conducted in the ocean depend in part on the distance from shore and in part on the particular strata of ocean space utilized (see Figure 3-1). For purposes of this analysis, OTEC installations will be considered as utilizing only the surface, water column, and seabed among ocean strata, excluding the atmosphere and the subsoil, although it is recognized that some superstructure will exist above sea level and that anchoring may indeed require burrowing into the subsoil. For analytical purposes the affected strata will be classified in two areas: (1) the surface and water column, and (2) the seabed.

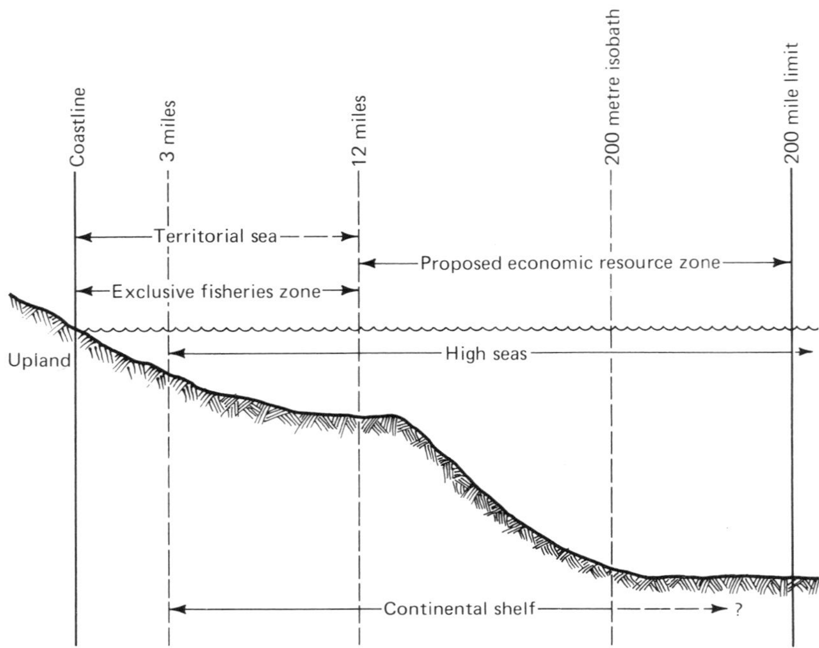

Figure 3-1. Profile of Continental Shelf (Vertical Scale Distorted).

The Surface and Water Column. Waters lying landward of the baseline[1] are classified as *internal waters*, over which the coastal state has exclusive and absolute jurisdiction with respect to all activities, subject to the general provisions of international law and specific international agreements to which the state is a party. The location of an OTEC device in internal waters would require the consent of the coastal state, and its operation would be subject to the laws and regulations promulgated by that state. International legal issues obviously would be minimal at such a site, but the absence of suitable water depths for the intake pipe in internal waters makes it unlikely that such a site would be chosen.

The *territorial sea* extends seaward from the baseline to a distance not yet subject to international agreement.[2] Most observers conclude that the maximum permissible breadth of the territorial sea under customary international law is 12 mi, although the United States still adheres to the policy of a 3-mi maximum breadth.[3] Within the territorial sea, the jurisdiction of the coastal state is near absolute, exceptions existing only for innocent passage and entry in distress.[4] For purposes of OTEC installations, then, deployment within the territorial sea would accord the coastal state exclusive competence to authorize construction of and regulate the facility (again subject to customary international law rules and to treaties to which that state might be a party).

Special contiguous zones extending beyond the seaward limit of the territorial sea are recognized for a variety of purposes in customary international law and by international agreement.[5] However, no special contiguous zones have been established or claimed for the purpose of authorizing the construction and regulation of artificial islands or installations for general economic purposes. Thus at present there is no special contiguous zone regime upon which a coastal state might predicate jurisdiction to construct or regulate OTEC devices.

Finally, the *high seas* consist of the waters and airspace seaward of the territorial sea, subject to the limitations of internationally recognized special contiguous zones.[6] The high seas may not be subjected to the sovereignty of any nation or appropriated to exclusive use by any nations or persons. Specified freedoms of the high seas include navigation, fishing, overflight, and the laying of submarine cables and pipelines, as well as "others which are recognized by the general principles of international law." Such freedoms are to be exercised "with reasonable regard to the interests of other States in their exercise of the freedom of the high seas."[7] An OTEC installation situated on the high seas would thus be beyond the *territorial* jurisdictional reach of a coastal state. Though no state may subject any part of the high seas to its sovereignty, states may regulate activities there either by virtue of the installations having been registered, licensed, or otherwise authorized by the state, or by virtue of laws and regulations applicable to its citizens wherever they may be found.

The Seabed. Since OTEC installations may be anchored to the seabed, brief consideration needs to be given to the various jurisdictional regimes applicable to those strata of ocean space.

The seabed underlying the *territorial sea* is, as noted above, subject to the exclusive jurisdiction of the coastal state,[8] and its consent would be required for any seabed use in that area.

The *continental shelf*, which, as an international legal (vis-à-vis geological) concept, begins at the seaward limit of the territorial sea and continues at least to the 200-m isobath and probably beyond to the edge of the continental margin structure,[9] is subject to the exclusive jurisdiction of the coastal state but only for purposes of exploring for and exploiting the natural resources of the area.[10] Consent of the coastal state is also required for the conduct of scientific research concerning the continental shelf and undertaken there.[11] Other uses of the shelf, however, are not subject to coastal state regulation. As a result, another nation could position an OTEC device on the high seas and use the coastal state's continental shelf for mooring purposes, provided that it did not interfere with the coastal state's exclusive right to exploit natural resources located in the same area.[12]

Beyond the seaward limit of the continental shelf lies the *deep ocean floor*, which is at present not subject to the jurisdiction of any nation. Pursuant to Resolution 2749 of the UN General Assembly, the mineral resources of the deep ocean floor constitute the "common heritage of mankind" and are to be exploited only pursuant to an international regime to be agreed upon. However, the majority of international legal scholars still consider deep seabed minerals as *res nullius*, with title to such resources vesting in him who first reduces them to his possession. Insofar as the anchoring of OTEC devices is concerned, there is neither a prohibition nor an express permission in international law, and this use of the seabed would therefore fall into the category of a reasonable use of the marine environment.

The upshot of this brief analysis of present international law is simply that if OTEC devices are to be located in internal waters or in the territorial sea, the coastal state will have exclusive jurisdiction with respect to such activities. On the other hand, if the OTEC device is to be situated on the high seas, then some basis other than territorial jurisdiction will have to be found in order to justify the deployment and regulation of such devices. Such other grounds might include the regulation of nationals, traditional admiralty law (where applicable), and special maritime jurisdiction, the latter predicated on the concept of reasonable use of the high seas.

The Developing Law of the Sea—The Third UN Conference on the Law of the Sea. Beginning in 1967 and continuing to the present date, the international community of nations has been actively engaged in negotiating a wide range of law-of-the-sea issues with a view toward adopting a timely, comprehensive, and widely accepted treaty on the subject.[13] Two substantive sessions of the Third United Nations Conference on the Law of the Sea have been held, one in June-August 1974, in Caracas, Venezuela, and another in March-May 1975, in Geneva.[14] Neither session resulted in adoption of treaty articles, and it appears at this time that unilateral actions may soon undermine the ability of the conference to develop international law through the comprehensive-treaty approach. If the conference should fail and as a result no international agreement relating to deployment of OTEC or similar devices is adopted, the future law governing such uses of the ocean will develop either on a customary law basis—in response to national claims to make particular uses of ocean space—or as a result of a multilateral treaty devoted either to OTEC devices or to offshore installations in general. Considering the possibility of failure of the conference, it would appear prudent for the United States to begin consultations with other nations possessing OTEC technology or indicating an interest in utilization of this form of energy conversion, to determine their interest in negotiating such an agreement.

Even though a global law-of-the-sea treaty may not be achieved in the near future, the law-of-the-sea debates both in the UN Seabed Committee and in the conference have made clear the aspirations and plans of most of the world's nations with respect to the law of the sea. One can derive from that record some estimate of future actions in the event of a conference failure.

Four developments in the conference have potential effects related to OTEC deployment. They are the concept of the economic resource zone, the notion of "common heritage" for deep seabed minerals, the possible regulation of scientific research throughout the world ocean, and the archipelagic concept.

Economic Resource Zone. The most important aspect of the negotiations for OTEC devices is the concept of an economic resource zone. Virtually every coastal nation in the world supports the concept, in one form or another, of coastal state jurisdiction over living and nonliving resources in the seabed and waters adjacent to its coast to a substantial distance from the baseline, usually specified at 200 mi.[15] At the conclusion of the Geneva session of the conference in May 1975, the chairmen of the three main conference committees produced unique treaty texts which, in combined form, are known as the "Informal Single Negotiating Text."[16] Although not representing agreement on the issues, the Negotiating Text does provide a basis for future negotiations. The provisions of Part 2, however, are said by some to represent, by and large, the basis for ultimate agreement.[17] Article 45 of Part 2 provides:

> 1. In an area beyond and adjacent to its territorial sea, described as the exclusive economic zone, the coastal State has:
> (a) sovereign rights for the purpose of exploring and exploiting, conserving and managing the natural resources, whether renewable or non-renewable, of the bed and subsoil and the superjacent waters;
> (b) exclusive rights and jurisdiction with regard to the establishment and use of artificial islands, installations and structures;
> (c) exclusive jurisdiction with regard to:
> (i) other activities for the economic exploitation and exploration of the zone, such as the *production of energy from the water*, currents winds; and
> (ii) scientific research;
> (d) jurisdiction with regard to the preservation of the marine environment, including pollution control and abatement;
> (e) other rights and duties provided for in the present Convention.
> 2. In exercising its rights and performing its duties under the present Convention in the exclusive economic zone, the coastal State shall have due regard to the rights and duties of other States. . . .[18] [Emphasis added]

The origin of the italicized language, seemingly applicable to OTEC devices, is not known at the present time. However, any agreement containing such a provision could easily be interpreted to give the coastal state exclusive jurisdiction with respect to OTEC and similar devices situated in its economic zone.

As noted above, if agreement to this effect is not reached by the conference, nonetheless it probably reflects the position which most coastal states would take on a unilateral basis, with the result that such a provision will likely become a rule of customary international law. Another possibility is the adoption by multilateral treaty of a similar provision for application on a regional basis. In such an instance it might be agreed that any state in the region could make such a use of the semienclosed sea area beyond territorial water limits, but the more likely outcome is coastal state exclusivity.

Common Heritage Concept for Deep Ocean Floor Wealth. Another of the significant aspects of the law-of-the-sea negotiations and the Third Conference, insofar as OTEC is concerned, was the adoption of UN General Assembly Resolution 2749 declaring the mineral wealth of the deep seabed beyond the limits of national jurisdiction to be the "common heritage of mankind."[19] This concept does not extend to the superjacent waters, nor even explicitly to minerals in solution, and it certainly does not extend to the sun's heat stored in those waters. Nonetheless, when one views the "philosophy of control" of the developing countries in pressing for the common-heritage concept, it does not take great imagination to envision the LDCs seeking to include energy and other resources in that area if it were politically feasible to do so.

Developed countries have fought hard to limit the scope of any international seabed authority to the regulation of deep seabed mining activities. Nonetheless, developing countries have pressed on occasion for "ocean space institutions" which would have jurisdiction over all activities taking place beyond the limits of national jurisdiction. This does not seem to be a likely outcome of the conference; it would certainly seem to be an unlikely development of customary international law through state practice in view of the developed nations' opposition. Nonetheless, the political environment is such that any use made of high seas areas beyond the economic zone will be viewed as a potential object for internationalization by the majority of developing nations.

Regulation of Scientific Research. Whether the conference produces an agreement or not, it seems likely that coastal states will assert authority to extend their rights with respect to scientific research on the continental shelf to the waters above the shelf. As noted above,[20] consent is presently required for the conduct of scientific research on the continental shelf of another nation. Many nations have argued forcefully that jurisdiction over scientific research in the waters above the shelf, as well as over pollution and navigation, is necessary if they are to manage effectively the exploitation of natural resources accorded to them by the economic-zone concept. Another question, which has not received substantial attention yet, is whether claims to regulate research in the water column would be extended beyond 200 mi where the coastal state's jurisdiction over the continental shelf exceeded that limit.

The provisions of the Negotiating Text concerning marine scientific research provide explicit rights and duties for both oceanographers and coastal states. The text also distinguishes between "fundamental" research and that related to resources of the economic zone in determining the degree of coastal state discretion with respect to such activities.[21] Should the conference fail, the claimed authority of coastal states over scientific research in the economic zone is apt to be more comprehensive—most likely a total "consent" regime in which permission must be secured before any scientific activities can be conducted in the economic zone. This could, of course, be a further basis for jurisdiction with respect to experimental OTEC operations within 200 mi of the coast.

With respect to the water column beyond an economic zone, traditionally scientific research has been regarded as completely free. However, in the law-of-the-sea negotiations, the economically underdeveloped nations, as part of their previously mentioned drive for control over ocean resources beyond the economic zone, have attempted to expand the competence of the nascent international seabed agency beyond mere regulation of seabed mining. The deep seabed mining articles of the Informal Single Negotiating Text are *not*—in contrast to the provisions of Parts 2 and 3—negotiated provisions reflecting an outline of potential agreement. Rather Part 1 (seabed mining) is simply a reiteration of the extreme position taken by the underdeveloped countries. Nevertheless, the provision on scientific research is illuminating:

1. Scientific research provided for in this Convention shall be carried out exclusively for peaceful purposes and for the benefit of mankind as a whole. *The [Seabed] Authority shall be the center for harmonizing and co-ordinating scientific research.*

2. The Authority may itself conduct scientific research and may enter into agreements for that purpose. . . .[22] [Emphasis added]

Unlike the provisions on scientific research in the economic zone, the Negotiating Text provisions of Part 1 do not reflect either the likely outcome of a successful conference or a likely development of customary international law. Nonetheless, this attitude toward scientific research in the ocean must be viewed as part of the political environment in which OTEC devices could be deployed.

The Archipelagic Concept. One final aspect of the law-of-the-sea negotiations which could affect deployment of OTEC devices is the developing principle of control over interisland waters by archipelagic nations. Under traditional international law, the "baseline" from which the seaward limit of offshore zones (e.g., territorial waters or fisheries zones) was measured, was limited to the low-water line along the mainland or islands of a nation, with exceptions for deeply indented coastlines, fringes of islands, bays, harbor works, roadsteads, islands, low-tide elevations, and river mouths. Even these exceptions did not allow archipelagic nations such as the Philippines or Indonesia to designate all interisland ocean areas as territorial or internal waters. The essence of the developing archipelagic concept is contained in the Negotiating Text:

An archipelagic State may draw straight baselines joining the outermost points of the outermost islands and drying reefs of the archipelago provided that such baselines enclose the main islands and an area in which the ratio of the area of the water to the area of the land, including atolls, is between one-to-one and nine-to-one.[23]

There are other limitations on the concept, such as the maximum length of such lines, but the principle is clear: archipelagic nations are likely, whether by agreement or by development of customary law initiated by unilateral claim, to assert henceforth a greater degree of control over their interisland waters. This means, of course, greater—probably exclusive—control over the deployment of OTEC devices in such waters.

The Impact of Unilateralism. Even as the law-of-the-sea negotiations were proceeding, many nations took steps which they regarded as necessary to protect their vital economic and security interests in the ocean. Canada, for example, promulgated a 100-mi pollution zone, ostensibly to protect its fragile Arctic environment.[24] Many developing nations also extended their fisheries jurisdiction to substantial distances from the coast. Iceland's extension of its fishing zone from 12 to 50 mi in September 1972 precipitated another round in the North Atlantic "cod war."[25] Many other examples could be cited.

Often it is assumed that the conference has had a dampening influence on unilateralism. That is, as long as the negotiations held out the promise of a timely, widely accepted, and comprehensive treaty which would protect all nations' interests in the ocean, there was less motivation to "go it alone." There is some evidence to support this theory since unilateral actions have not been as extensive as they might have been in a situation where nations saw no other alternative to the protection of their ocean-related interests. Also it is often assumed that should the conference break down or otherwise fail to produce a treaty in a timely fashion, there would occur a rash of unilateral actions as nations moved to protect their interests in the only way left to them. Although such scenarios are probably overstated, it is likely that coastal states will assert substantial competence over adjacent ocean waters to a distance of 200 mi from the coast. It is virtually certain that activities such as deployment of OTEC devices would be subject to such claims.

With respect to ocean areas beyond these assertions of jurisdiction, unilateral actions will, by and large, take the form of the conduct of activities by one or another nations under their flags or regulations. For example, the United States recently enacted legislation authorizing the construction of deep-water port facilities beyond the limit of the territorial sea.[26] In doing so, it based its authority on the theory of reasonable use of the high seas,[27] and to date there have been no protests or other indications of opposition to this action. Since OTEC installations situated beyond any reach of national jurisdiction will take the form of an overt act by the nation supporting such activities, an investigation of the reasonable-use concept as a basis for deploying and operating such devices is warranted. This analysis applies at present to all waters beyond the territorial sea. If the 200-mi economic resource zone concept is adopted, however, through either written agreement or customary law development, and if it provides for exclusive coastal state control over OTEC deployment, then the following discussion would be applicable to only the area beyond the economic zone.

The Reasonable-Use Theory

International law is a dynamic process rather than a static body of rules. One method for changing old rules or initiating new ones is for a state or group of states to begin to act in a new manner, desirable to them, or to claim the legal right to so act. Responses from other nations soon make clear whether—and if so, under what conditions—a new course of action will be tolerated. This process of adjustment of attitudes through claims and responses, counterclaims, and counterresponses leads, over time, to the emergence of new normative standards of international conduct. The "reasonableness" of the initial claim, however, has particular consequences for the evolution of legal principles.

From an early date it became apparent that coastal state jurisdiction over a relatively narrow band of territorial waters was inadequate in some instances to protect perceived economic and security interests. The concept of freedom of the high seas did not permit the appropriation of areas beyond territorial waters, nor did it permit the exercise of national authority there save with respect to regulation of activities taking place on vessels. Two sets of events—the improvement in ship capabilities, which permitted smuggling activities to be centered well beyond the limit of the territorial sea, and the development of technologies in the twentieth century that permitted a wide range of new uses of the high seas—dictated that new rules be developed to authorize or regulate such activities.

The landmark judicial decision recognizing the interests of coastal states in areas beyond their narrow territorial waters is *Church v. Hubbart.*[28] That case involved the seizure of a vessel on the basis that it was involved in illicit trade with the Portuguese colony of Brazil. On the question of the legality of the seizure and the interests which might thereby be legitimately protected, the U.S. Supreme Court stated that a nation's power to secure itself from injury could be exercised beyond the limits of its territory. Most of the commentaries and developments which followed this decision were predicated on the *defensive* requirements of such action ("power to secure itself from injury"). Not until the twentieth century were there significant expressions of a "positive" or "authorizing" right to take action on the high seas which might be generally regarded as inconsistent with the nonappropriation system then prevailing. Nonetheless, the germ of the reasonable-use theory had been sown. For example, writing in 1927,

Judge Philip C. Jessup concluded with respect to the exercise of jurisdiction over vessels beyond territorial waters in cases of potential threats to a coastal state:

There seems, however, to be sufficient evidence of acquiescence in *reasonable claims* to warrant the assertion that a customary rule of international law has grown up under which such acts may be held legal if they meet the test of a *reasonableness.* The relative vagueness of this norm makes it necessary to state that a nation acts in such cases at its peril, its definitive vindication depending upon ultimate determination by some international tribunal.[29] [Emphasis added]

Likewise, writing in the 1959 edition of his work on the law of the sea, H.A. Smith comments:

If the general law, apart from exceptions created by special treaties, fails to give reasonable protection, then the law is defective and should be amended by common consent. The law of nations, which is neither enacted nor interpreted by any visable authority universally recognized, professes to be the application of *reason* to international conduct. From this it follows that any claim which is admittedly *reasonable* may fairly be presumed to be in accordance with law and the burden of proving that it is contrary to law should lie on the State which opposes the claim.[30] [Emphasis added]

In making his defense of nuclear weapons testing on the high seas, in the 1950s, Myres McDougal developed the reasonable-use theory further. In his major article on the subject, coauthored by Norbert Schlei, McDougal stated:

The technical prescriptions of the "freedom of the seas" on the one hand, and of "territorial seas," "contiguous zone," "jurisdiction," and so on, on the other, are not arbitrary, inelastic dogmas, but rather are highly flexible policy preferences invoked by decision-makers to record or justify whatever compromise or adjustment of competing claims they may reach in any particular controversy. And for all types of controversies the one test that is invariably applied by decision-makers is that simple and ubiquitous, but indispensable, standard of what, considering all relevant policies and all variables in context, is *reasonable* as between the parties.[31]

In their seminal work on the law of the sea, McDougal and William T. Burke qualify this concept by noting:

Because of certain contemporary misconceptions, it may be emphasized that this formulation does not mean that each state may itself unilaterally decide what is reasonable with respect to its exclusive claims and lawfully impose its decision upon the community. It is not the unilateral claim, but the acceptance by other states, even when manifested in reciprocal tolerances, which creates the expectations of uniformity and "rightness" in decision which we commonly call international law. . . . Thus, the end of claimed unilateral competence can only be irresolvable conflict. Multilateral concurrence, whether by customary consensus or by explicit undertaking, is indispensable to accommodation.[32]

Two major points of inquiry are raised by this analysis. First, is the reasonable-use theory a sufficient justification for asserting the right to implant OTEC devices initially, subject to the jurisdiction of the United States (or any other nation), in high seas areas beyond the jurisdictional reach of other states? Second, what kind of justifications or assertions would the United States (or any other nation) need to make in concert with the deployment of OTEC devices beyond national jurisdiction in order to make the case most effectively for legality of the action?

As to the first inquiry, a strong case can be made for an affirmative answer based upon the recent precedent of the deep-draft port legislation in the United States.[33] When considering legislation to authorize the construction of such facilities beyond the limit of the territorial sea, the National Security Council's Inter-Agency Law of the Sea Task Force evaluated a wide range of possible jurisdictional bases with respect to superports. The Task Force concluded that the reasonable-use argument was the preferable one, based in no small part on the desire to avoid the possible adverse effects of territorial claims on the current law-of-the-sea negotiations. Even though that constraint probably will not be a relevant factor at the time OTEC installations are deployed, the reasonable-use theory is probably the best basis upon which to predicate the deployment and regulation of such devices. In his testimony before the House Committee on Merchant Marine and Fisheries concerning the superport bill, John Norton Moore, chairman of the Task Force, stated:

> The question of reasonableness is determined by looking at all relevant factors. It would be necessary, for example, to ensure that deepwater port facilities did not unreasonably interfere with the high seas freedoms including navigation, fishing, laying submarine cables and pipelines and overflight. . . .
>
> [T]he United States would be exercising the international legal right to make a permissible use of the high seas. . . . This approach recognizes the vitality of international law and illustrates the wisdom of maintaining the flexible high seas doctrine over as broad an area as possible. Competitors for the same ocean area can be accommodated so long as they reasonably take into account each other's rights and interests.[34]

The second question is more difficult to analyze. Reasonable use, in the end, is simply a justification for the assertion of a unilateral claim to make a particular use of some portion of ocean space. The final judge of reasonableness is the international community, although that community will consider those factors urged by the claimant state in justifying its claim. According to Smith, some uses are "admittedly" reasonable, and in such cases states opposing the claim must prove that it is unlawful, not simply unreasonable. Further, some uses may be reasonable when considered in general, e.g., drilling offshore oil wells; but when considered in particular, e.g., drilling a well located in a recognized sealane, the use would be considered unlawful.[35] In either case, however, the claimant state must present a justification for its claim, for that is necessary in determining the reasonableness of the action.

In determining whether a use is reasonable, two factors must be considered: its reasonableness in the light of the interests of the acting state, and its reasonableness as to the interests of the international community as a whole. The paradigm case is the United States claim to resources of the continental shelf. In the Truman Proclamation of 1945,[36] the United States carefully listed its reasons for asserting jurisdiction over continental shelf mineral resources: the national and international need for new sources of petroleum; the practicability of extraction of such resources from the continental shelf; the need for jurisdiction in order to ensure the conservation and prudent utilization of the resources; the contingency of coastal cooperation and protection; the geological fact of the shelf constituting an extension of the continental land mass; the fact that some deposits formed a seaward extension of a pool lying within the territory of the coastal state; and the national security aspects of the conduct of mining activities off the United States coast by other nations.

The extension of jurisdiction obviously had important economic and security effects for the United States. At the same time, the proclamation limited the jurisdiction claimed to the resources of the seabed and subsoil, emphasizing that the superjacent waters would remain high seas. Thus the international community had a clear picture of the interests of the acting state, but it also had assurances that its interests in freedom of navigation would be protected. The international community knew that in actual practice the development of offshore drilling would result in some interference with navigation, but the advantage to each coastal state in being able to exploit fully the oil reserves to be found beneath its own continental shelf, coupled with the need for new supplies of petroleum and natural gas, apparently outweighed that disadvantage. The approval of the international community was seen in the complete lack of protest and in the speed with which other nations made similar pronouncements.

A quite similar approach might be successfully taken with respect to OTEC devices. By no means the only possible approach, the following itemization of justifications is intended only to indicate the *types* of considerations which ought to be addressed; and it presents, only by way of example, some arguments that could be adduced in support of OTEC deployment beyond limits of national jurisdiction.

First, the claiming state could make a strong case concerning its economic and security interests which would be served by such action. In this era of high-priced conventional fuels and the threat of control of energy supplies and raw materials by underdeveloped nations, such a case should not be difficult to make. It is, in short, the desire for energy independence which is already a major facet of United States energy and foreign policies.

Second, the claiming nation could make an argument based on alternatives to OTEC. It could attempt to show either that pursuit of alternative energy sources would be inferior to OTEC deployment or, more likely, that OTEC is an integral feature of a "mix" of energy development programs necessary to protect economic and security interests of the claiming nation.

Third, the claim could point out how the broader interests of the international community are protected *or* served by such action. For example, it could be noted that the legal right to free navigation on the high seas is not to be impaired although, as with the case of continental shelf installations, there will in fact be some minor interference with navigation. Provisions for appropriate safety zones and warning equipment could assist in ameliorating concern over this issue. Beyond this "negative" approach, however, the case could be made that all nations will ultimately benefit from the development of this new source of energy. Just as continental shelf oil and gas development greatly assisted the economies of many underdeveloped nations, so the perfection of OTEC devices could be of benefit to them, particularly those located in tropical or semitropical waters (which includes nearly all the underdeveloped world).

The sticking point in this process relates to the matter of *exclusivity* with respect to OTEC deployment. If the claim were simply limited to the right to use a particular area of the ocean for an OTEC installation, without asserting an *exclusive* right to a broader area, deployment is likely to meet with little protest (provided the proper justifications are made for the claim). On the other hand, if a significantly large area of the ocean is prohibited to other OTEC devices by the claim (because of the drain on heat caused by two devices in too-close proximity to each other) or to other uses in general, then the issue has escalated to a new plane.

There are no precedents for successful claims of exclusive use of ocean space beyond limits of national jurisdiction. Indeed, such claims are uniformly resisted on the basis that maximum use of the ocean can be made only by fostering the policy of inclusive use by all nations. Therefore it will require straining the reasonable-use theory to the breaking point to argue that an exclusive right is required. Nonetheless, such an argument has a sound base in reason: namely, unless certain distances are maintained between OTEC units, they will become inoperative and none of the benefits for either the claiming state or the international community at large will be obtainable. The real problem is in allocation. Should a particularly valuable piece of ocean for OTEC devices be accorded on a first-come, first-served basis, or should there be some formal allocation procedure which would take into consideration factors other than technological advancement? The same issue is being dealt with at the present time in the deep seabed mining controversy, and the outcome of that debate will have profound implications for the OTEC deployment problem. More attention will be given to this issue in the next section.

Suffice it to say that the theory of reasonable use, which is simply a short term for "initiation of customary international law development through unilateral action," does provide a basis for deploying OTEC devices beyond limits of national jurisdiction. And such a claim could be backed by justifications demonstrating (1) the economic and security importance to the coastal state, (2) the lack of alternatives or the overriding importance of the OTEC alternative, and (3) the broader interests of the international community protected or served by deployment.

Ownership of Ocean Thermal Energy

Another major international legal issue involved in OTEC installations concerns the ownership of the resource being exploited and the rights, if any, of the owner to extract rents or royalties for use of it. Although it may seem farfetched to some to discuss "ownership" of the energy contained in emissions from the sun, recent international developments, particularly with respect to natural resources, indicate that an assessment of this issue may be of substantial significance.

The analysis will be developed by first identifying several analogous situations and then pointing out where among them OTEC seems best to fit.

Analogies

The purpose of this section is to discuss how "ownership" concepts are applied or have been proposed to be applied to a range of other ocean resources.

Fish. Within the territorial sea (or, more recently, within claimed exclusive fishing zones) the coastal state traditionally has asserted rights of "ownership" of living marine resources. Within the concept of ownership, however, the fish have generally been regarded as *res nullius*, with title to fish being vested in the fisherman who first reduces them to his possession (provided he complies with regulations, if any, pertaining to the particular species involved). *Res nullius* literally means "the property of nobody," and in the sense used here it represents a thing which has no owner because it has never been appropriated by any person, yet is of the class of things which must be reduced to possession and consumed in order to be enjoyed by people. The seeming conflict between state ownership and a *res nullius* characterization was addressed by the U.S. Supreme Court in the case of *Toomer v. Witsell:*

> The whole ownership [of fish] theory . . . is now generally regarded as but a fiction expressive in legal shorthand of the importance to its people that a state have power to preserve and regulate the exploitation of an important resource.[37]

On the high seas, beyond the reach of any territorial sea or fishing zone, fish are clearly considered as *res nullius* because of the customary and conventional international law principle of freedom of fishing on the high seas. Of course, nations retain jurisdiction over their nationals and can thus regulate their fishing activities on the high seas, just as they can enter into international agreements with other nations to limit or regulate their fishing activities on the high seas. But under this rule of capture, or principle of "open access," no nation or fisherman may unilaterally appropriate areas of the high seas to his exclusive use. The applicable principle becomes simply that title to fish is vested in him who first reduces them to his possession. It should be noted, however, that pursuant to Article 2 of the Convention on the High Seas, fishermen must exercise this freedom of fishing with reasonable regard to one another and to other users of the marine environment. Except for regulations imposed by international agreement among affected states, or by nations with respect to their own citizens, there are no restrictions on the right to extract fish from the high seas.

Recent studies have indicated that this philosophy with respect to high-seas fisheries may have undesirable biological and economic effects for the fishing industry.[38] Fish, as we now know, do not exist in unlimited quantities, and overfishing may result in reduction of stock size to a point where the resource can no longer be enjoyed by humans. Likewise, the open-access principle inevitably leads to overcapitalization and concomitant gear congestion and economic waste. Thus the current trend is away from treating fish as *res nullius* and toward creating jurisdictional bases for management (or "property rights") so that rational management systems can be effectuated.

Is ocean thermal energy analogous to fish? Certainly under the classical definition of *res nullius* it would seem so. Solar energy is still the property of no one. It is a thing which has no owner because it has never been appropriated by any person. Yet it is of the class of things which (at least in the form sought here)[39] must be reduced to possession and consumed if it is to be enjoyed by humans. The next question, then, is whether ocean thermal energy possesses attributes so similar to fish that we must anticipate the trend from *res nullius* to property rights for OTEC. My own estimation is that we need not, at least for the foreseeable future. The amount of solar energy falling between approximately 10°N and 10°S latitude exceeds current and projected world demand by many orders of magnitude, and astrophysicists tell us that the sun can be expected to continue to generate energy for several billion more years. Nonetheless, preliminary studies indicate that in order to operate efficiently, OTEC devices require a thermal differential on the order of 20°C occurring at a water depth of about 1500 m. This reduces the number of available sites and leads one to suspect that at some future date there might in fact be competition for the same locations. Until that situation presents itself, however, it would seem appropriate to avoid the bureaucratic and economic burdens associated with allocation of property rights, and to consider that ocean thermal energy is in fact analogous to fish at the point in time when demand did not result in either overfishing or overcapitalization of the fishery.

Ocean Buoyancy; Navigation. The "resource" of the ocean which makes it useful for maritime commerce is its buoyancy. It was recognized from an early date—emphasized in Grotius' treatises on the law of the sea—that the use of the buoyant property of the ocean by one vessel in no way detracts from its subsequent similar use by other vessels. Thus the use of the sea for navigation could be characterized as *res communes* since it was not necessary to appropriate the resource in order to enjoy the benefits of its use. *Res communes* literally means "things common to all," and in the sense used here it is that class of things which are used and enjoyed by everyone in parts but which can never be exclusively acquired as a whole (traditional examples are light and air). This being the case, no nation or person has the right to appropriate any area of the ocean for its buoyant properties in order to secure a monopoly on the right of trade. As the quantity of navigational use of the ocean increased, it became necessary in limited instances to regulate passage, particularly in narrow confines such as the Straits of Dover and at the entrances to major ports. For the high seas far from any land masses, however, the concept of navigation as *res communes* still prevails.

Is ocean thermal energy analogous to navigational use of "high seas"? It would appear not, because of the critical difference with respect to the necessity for appropriation. Ocean thermal energy must be appropriated and converted before it is useful to people; the buoyant property of the ocean requires no such appropriation for its utility (save for the "possession" of the area in which the vessel is located at any instant—a phenomenon which has given rise to "rules of the road" and other navigational safety institutions). The better view is to consider ocean thermal energy, as noted in the preceding section, as *res nullius.*

There are lessons to be derived from the navigational experience, however: mainly, congestion will occur at some future date, and some rules governing allocation of space or rights of passage may need to be developed. Further, if OTEC installations are to utilize the mobile mode, then insofar as their navigational activities are concerned, they would be analogized nicely to ships. This is, of course, a distinct issue from the extraction of energy from the ocean.

Manganese Nodules. Until the adoption of Resolution 2749 by the UN General Assembly in 1970,[40] it seemed clear that mineral resources on the seabed beyond the limits of continental shelf jurisdiction were *res nullius*, subject to title vesting in him who first reduced them to his possession. That resolution, however, characterized such resources as the "common heritage of mankind," subject to exploitation only pursuant to an international regime to be agreed upon. Since that international regime has not been forthcoming, some nations assert that Resolution 2749 has a "moratorium" effect and that further mining activities should be terminated. Since the overwhelming support for Resolution 2749 was expressed more in order to produce progress in the negotiations than to classify the resources in question, the majority view would appear to be that the resources still retain their quality of *res nullius.* For experimental work, and probably for initial development work, the concept of *res nullius* would appear to be adequate. However, as recent events have shown, most operators desire exclusivity of tenure. Those entities (whether lending institutions or boards of directors) committing funds to deep seabed mining enterprises wish to ensure that their capital investment will be protected through the right to remain in a particular location and mine it to the exclusion of all others. The *res nullius* concept does not, of course, include the concept of exclusivity. It was for this reason that Deepsea Ventures filed its claim to exclusive mining rights with respect to a site in the Pacific Ocean in November 1974.[41] In the event the conference fails to produce a seabed mining regime through a comprehensive and widely accepted international agreement, it appears likely that those nations possessing the requisite technology will enact legislation which can be coordinated in practice in order to prevent claim jumping and to allow exclusivity through the process of mutual recognition of claims and the regulation of their citizens rather than through territorial claims.[42] This evolution from *res nullius* to "quasi-property rights" is necessitated by the economics of the industry.

Is ocean thermal energy analogous to manganese nodules? Insofar as their present *res nullius* status is concerned, it would appear so. A critical distinction, however, lies in the nature of the respective industries–does OTEC require exclusivity of tenure in order to be economically viable? Could an OTEC installation continue to produce the requisite amount of energy if other OTEC devices were to "poach" on a position of highly favorable thermal differential, shallow depth, and calm environment? The manganese nodule mining industry has made a strong case for the necessity of exclusive tenure. It would appear that more needs to be known about the operation of OTEC devices before a clear answer can be given to the questions posed above.

If the extraction of heat and the concomitant discharge of cold water at or near the surface would not affect the operation of another OTEC device (and vice versa) located in close proximity, then there would seem little reason to provide a property right basis from which exclusive tenure could be derived. More general principles, such as the notion of "reasonable regard to the interests of other States in their exercise of the freedom of the high seas," would appear appropriate. On the other hand, if close proximity results in inefficiencies which would reduce the output of OTEC installations and thus discourage the requisite flow of capital and technology, some form of exclusive tenure will have to be developed. Taking a cue from the manganese nodule situation, it would seem preferable to establish such rights first on a nationality basis, through domestic legislation in those states possessing OTEC technology, followed by a concordant practice among those nations. The international agreement approach would not seem particularly fruitful, if the conference is any indication of developing nations' attitudes.

Conclusion

If both legal theory and state practice with respect to resources of the high seas are considered, it would seem that ocean thermal energy is at present *res nullius.* There is no indication that such energy is so potentially exhaustible that property or quasi-property rights need to be developed now. Appropriation is required for beneficial use of ocean thermal energy, yet territorial claims do not appear warranted at this stage. Should congestion become a problem, the better solution would appear to be coordinated domestic legislation among nations possessing OTEC technology.

Within near-shore coastal areas, it seems clear that any economic use of waters will come under exclusive coastal state jurisdiction. The concept of the economic resource zone has already been discussed, and it need only be pointed out here that the provisions of the Negotiating Text would clearly seem to include the deployment of OTEC devices. Thus, within 200 mi of the coast, one can—probably within 2 years from the present date—assume that the consent of the coastal state will be required for the deployment and operation of OTEC installations. Beyond the 200-mi limit, the deployment of OTEC devices should be permissible as a reasonable use of the high seas.

The only major obstacle to the validity of these conclusions is that some developing nations may seize upon the issue of ocean thermal energy, as they have seized upon the issue of deep-sea mining, for political purposes. One can envision an attempt to declare ocean thermal energy as the common heritage of all human beings on the planet or some equivalent characterization. If that should occur, the scenario which has evolved with respect to deep seabed mining could be expected to repeat for ocean thermal energy. This would probably involve demands for a moratorium on OTEC operations. There could follow an attempt to establish a bureaucracy to license and regulate the deployment of OTEC installations beyond the limits of national jurisdiction. Such negotiations would likely become bogged down in philosophical differences, just as the debates over deep seabed mining have, and the likely outcome is that there would be no internationally agreed-upon regime. There would then exist two competing practices: (1) that of the developed nations which would claim the legal right to deploy OTEC devices on a reasonable-use basis, and (2) that of the developing nations which would assert that any extraction of solar energy from the ocean was a violation of international law until such time as they had developed an appropriate mechanism for licensing and regulating such activities.

Jurisdiction to Protect OTEC Installations

It is unfortunate—but nonetheless a sign of our times—that any significant effort should be expended on an analysis of the necessity for protecting OTEC installations from accidental and deliberate interference. We live in an age of increasing use of the ocean and increasing competition for ocean space and its resources. We also live in an age of conflict and terrorism. These two facts make it necessary to assess the legal framework for protecting OTEC installations from other uses of the sea, and vice versa, and for protecting such installations from deliberate injury or interference.

The Safety Zone Concept

As the uses of ocean space began to intensify, it became clear that uses requiring the placement of fixed installations in the ocean threatened interference with traditional navigational use of the high seas. The framers of the Convention on the Continental Shelf recognized this conflict and provided that a coastal state might establish safety zones around installations designed for the exploitation of continental shelf resources and in such zones take "measures necessary for their protection."[43] These safety zones are limited to a radius of 500 m around the installations, and ships of all nationalities are required to respect them.[44] The convention further provides that neither the installations nor their safety zones may be established where "interference may be caused in the use of recognized sea lanes essential to international navigation."[45] Finally, coastal states are obliged to undertake in safety zones "all appropriate measures for the protection of the living resources of the sea from harmful agents."[46]

This rather conservative approach sufficed in most instances. However, in situations where offshore installations are extensive, such as in the Gulf of Mexico off the Louisiana coast, other solutions have been sought. In that instance, shipping safety fairways were devised. Such fairways consist of areas which the U.S. Army Corps of Engineers specifies are free of installations and in which none are expected to be constructed in the future.[47] Such lanes are not mandatory, but are simply indicated on nautical charts as areas in which one would not expect to encounter artificial structures. Delimitation of these areas on maps has been ineffective since in the absence of a requirement to navigate within them, ships continue to take the more economical (at least in terms of fuel consumption) routes through the heart of the offshore oil fields.

The efforts of the Intergovernmental Maritime Consultative Organization (IMCO) to establish traffic separation lanes in the Dover Straits and at access points to ports is well known and need not be elaborated here. What is important to note is that an international body has been authorized to restrict navigation to particular areas because of traffic congestion and multiple-use problems.

Is the traditional safety zone concept sufficient in the case of OTEC installations? Such installations are relatively small—at least as presently planned—and would seem to be similar in some respects to offshore-resources exploitation installations. If experience dictated that 500-m-radius safety zones have been adequate for offshore oil installations, there is no reason to establish a more elaborate system—at least for purposes of conflicts with navigation. On the other hand, if OTEC deployment becomes as intense in a particular area as in the case of offshore oil installations in the Gulf of Mexico, some other approach may have to be taken, hopefully one more effective than the system of shipping safety fairways utilized in the Gulf of Mexico.

Some tentative designs for OTEC devices envision mooring such structures to the seabed at great depths with the result that, depending on prevailing wind and current conditions, the device could meander over quite a large area of ocean space. This would render impracticable the designation of such devices on nautical charts, since their specific location at any time would be uncertain. It would dictate more emphasis on audible, visual, or electronic warning devices located on the device itself, perhaps coupled with an indication on nautical charts of the "watch circle" area somewhere within which the device would be located.

Still another approach would be to develop an overall multiple-use plan for coastal areas in which uses would be cleared and systems designed to ameliorate conflict between competing interests. This requires a fairly sophisticated approach to the problem which in turn requires a fairly sophisticated approach to multiple-use problems in the ocean. Because the issue of multiple use is addressed in more detail in another chapter,[48] no further analysis of it will be made here.

Problems of Deliberate Interference

Recent incidents in the North Sea have triggered concern for the safety of offshore installations. In at least one instance, Soviet intelligence-gathering vessels have harassed an oil and gas installation on the United Kingdom's continental shelf.[49] This threat has resulted in a reconsideration of methods for protecting offshore installations by both the United Kingdom and NATO.[50] At present, neither the United Kingdom nor NATO has published any specific plan to deal with the problem of deliberate interference, but they are both expected to do so in the near future. The problem is compounded by the possibility of terrorist activities with respect to offshore installations. Accordingly, it is apparent that some system will have to be devised to permit effective enforcement of security zones around all offshore installations which are threatened with potential harm.

The jurisdictional base for such protective measures is not altogether clear. Customary international law has always recognized a right of self-defense in the face of attack from another nation. There is little doubt that if such an attack occurred against an offshore installation, it would be considered an attack against the nation authorizing or licensing such installation. The right of self-defense is, of course, preserved in the UN Charter.

The vacuum in present international law and policy concerns preventive measures. For example, a nation licensing an OTEC installation within its economic zone might attempt to require vessels operating within some fixed distance of such a device, say 5 mi, to subject themselves to search on the high seas. The only good analogy occurs in the field of offshore installations authorized for continental shelf exploration and exploitation. Article 5(2) of the Convention on the Continental Shelf authorizes coastal states to "establish safety zones around such installations and devices and to take in those zones measures necessary for their protection." Presumably this would include barring vessels from navigating within the zone, but because the maximum size of such zones is 500 m in radius, little protection against deliberate interference is afforded. Indeed, the original concern with such zones was not deliberate interference but navigational safety and avoidance of fire hazard.

Pursuant to the Convention on the Continental Shelf, the United States enacted the Outer Continental Shelf Lands Act[51] which provides that:

> The Constitution and laws and civil and political jurisdiction of the United States are extended to the subsoil and seabed of the outer Continental Shelf and to all artificial islands and fixed structures which may be erected thereon for the purposes of exploring for, developing, removing, and transporting resources therefrom. . . .[52]
>
> The head of the Department in which the Coast Guard is operating shall have authority to promulgate and enforce such reasonable regulations with respect to lights and other warning devices . . . on the islands and structures . . . or on the waters adjacent thereto, as he may deem necessary.[53]

Enactment of this law met with no protest from other members of the international community. Should the same approach be taken for OTEC devices, no protest is to be expected, provided that the encroachment on inclusive uses of the high seas is not significant. Assertion of a right to board and inspect vessels navigating within 5 mi of an OTEC device, for example, would unquestionably meet with substantial protest, and such a claim is likely to be rejected.

Another method of securing jurisdiction would be to emulate the approach taken in the European Agreement for the Prevention of Broadcasts Transmitted from Stations Outside National Territories.[54] Rather than asserting territorial jurisdiction with respect to operators of "pirate" radio broadcasting stations situated on the high seas off the coast of European nations, the concerned nations agreed in that treaty to individually enact prohibitory laws concerning their nationals and to prosecute such nationals for illegal activities. The difficulty with such an approach for OTEC devices is in securing such agreement from virtually every nation in the world. Further, this would offer little guarantee against terrorist organizations who, by and large, have been able to avoid punishment by the nations from which they operate.

Perhaps the best approach would be to secure international acquiescence—whether through treaty or customary law development—in coastal or flag state protection of such devices by authorizing them to exercise normal police powers, in conformity with domestic laws, to prevent deliberate interference with such facilities. Exactly what specific powers would need to be accorded the coastal or flag state will depend in large part on the ultimate physical characteristics of OTEC devices and the places where they are sited.

Miscellaneous Issues

Inspection Pursuant to Seabed Denuclearization Treaty

The seabed denuclearization treaty prohibits the deployment of nuclear weapons or other weapons of mass destruction on the seabed beyond 12 mi from the coast.[55] Article 3 of that treaty provides in part that:

> In order to promote the objectives of and ensure compliance with the provisions of this Treaty, each State Party to the Treaty shall have the right to verify through observation the activities of other States Parties to the Treaty on the sea-bed and the ocean floor and in the subsoil thereof . . . provided that observation does not interfere with such activities.[56]

It is arguable that if a nation had reasonable doubts about the violation of the treaty by a state party in the form of a purported OTEC installation, it could require observation of the facility in order to assure itself that no treaty violation was in fact occurring. The inspection provisions of the treaty have never been invoked; therefore it is difficult to assess the logic of such a claim. Presumably some sort of *prima facie* evidence would have to be provided; and considering the nature of OTEC plants (their operation is unrelated to the seabed), this would seem to be quite difficult to achieve.

With respect to OTEC facilities located above the legally defined continental shelf, a question has arisen concerning whether consent of the coastal state is necessary to invoke the inspection provisions of the treaty. Yugoslavia submitted an interpretive statement following its ratification of the treaty to the effect that a state exercising the right of inspection is obligated to notify a coastal state in advance if the observations are to be carried out in waters above the latter's continental shelf. The United States has objected to this interpretation on the basis that it is contrary to the existing law of the sea and that as a reservation to the treaty it would be unacceptable because it is incompatible with the object and purpose of the treaty.[57]

In any case, the application of Article 2 would not appear to involve any potential interference with the operation of OTEC plants, particularly in light of the provisions that the inspection or observation shall not result in any interference with the activities being undertaken. Nonetheless, it would seem appropriate to develop a theory of interpretation of the inspection provisions such that any request for inspection could be met either with a well-reasoned argument requiring the showing of a *prima facie* case or with a contingency plan to facilitate the inspection if that were the desired course of action.

Flags of Convenience

An initial question arises concerning the legal nature of an OTEC installation (regardless of whether it is fixed or mobile): Is it a vessel subject to maritime and admiralty law? This issue is beyond the scope of this chapter and is being dealt with in another chapter in this book.[58] It should be noted, however, that the international law of the sea does recognize the right of all states to register vessels for navigation on the high seas.[59] Presumably, all OTEC installations would be authorized by some state and could be said to possess the "flag" of that state, whether or not such installations were assimilated to vessels for purposes of admiralty and maritime law. That being the case, an issue is raised concerning "flags of convenience."

As is well known, some vessel owners register their vessels in states where labor, taxation, and environmental standards are low or poorly enforced in order to secure maximum economic benefit from the shipping enterprise. The First UN Law of the Sea Conference attempted to deal with this phenomenon by providing, in Article 5 of the Convention on the High Seas, that:

> Each State shall fix the conditions for the grant of its nationality to ships, for the registration of ships in its territory, and for the right to fly its flag. Ships have the nationality of the State whose flag they are entitled to fly. There must exist a genuine link between the State and the ship; in particular, the State must effectively exercise its jurisdiction and control in administrative, technical and social matters over ships flying its flag.[60]

The literature on the subject of flags of convenience is abundant, and no attempt will be made here to analyze the problems involved.[61] Suffice it to say that one must first make an argument by analogy in order to subject OTEC installations to a "genuine link" test; then the question of the substantive content of the genuine-link test must be ascertained. As noted above, most of these issues are more germane to the subject of responsibility and liability and are dealt with at greater length in Chapter 8.

Finally, questions of sovereign immunity may be raised in situations where governments deploy OTEC installations and regard government ownership as entitling such installations to sovereign immunity. Again, the literature on the question of sovereign immunity for state-owned ships in commercial service is abundant, and the issue will not be discussed in any detail here.[62] The same two problems noted above—reasoning by analogy and determination of substantive content of the rule—would be applicable to this analysis.

Cables

The Convention on the High Seas recognizes in Article 2 the freedom to lay submarine cables on the seabed underlying the high seas. However, where the seabed is classified as continental shelf, other provisions of that treaty are also relevant. Article 26 provides:

> 2. Subject to its right to take reasonable measures for the exploration of the continental shelf and the exploitation of its natural resources, the coastal State may not impede the laying or maintenance of such cables or pipelines.
> 3. When laying such cables or pipelines the State in question shall pay due regard to cables or pipelines already in position on the seabed. In particular, possibilities of repairing existing cables or pipelines shall not be prejudiced.[63]

A similar provision appears in Article 4 of the Convention on the Continental Shelf. Obviously, then, the operators of OTEC installations will be required both to obtain the consent of coastal states for the laying of transmission cables providing electricity to the shore and to comply with appropriate regulations concerning maintenance and liability of such cables. The political feasibility or desirability of entering into such arrangements may dictate selection of alternative locations as well as alternative end products from OTEC installations. Further, the mooring of OTEC devices has a potential for interference with existing cables and pipelines, and conformance with laws and regulations of the coastal state would therefore be required even for mooring.

Notes

1. For parties thereto, the baseline is determined by the rules set forth in Articles 3 through 11 and 13 of the Convention on the Territorial Sea and the Contiguous Zone [*done* April 29, 1958, 15 U.S.T. 1606 (1964), T.I.A.S. No. 5639, 516 U.N.T.S. 205, *in force* Sept. 10, 1964]. Normally, the baseline is the

low-water line along the coast of the mainland or islands, but exceptions exist for deeply indented coastlines, fringes of islands, bays, harbor works, roadsteads, islands, low-tide elevations, and river mouths.

2. Neither the first nor the second UN conference on the law of the sea was able to reach agreement on the breadth of the territorial sea. In the third conference, now underway, a consensus seems to have developed in favor of a 12-mi breadth, but the meeting has not yet produced a written agreement. There has been no international adjudication in which the question has been decided.

3. According to 1974 data issued by the U.S. Department of State, 43 nations claim territorial sea breadths less than 12 mi, 52 claim exactly 12 mi, and another 20 claim in excess of 12.

4. Articles 1 and 2 of the Convention on the Territorial Sea and the Contiguous Zone, *supra* note 1, provide:

> The sovereignty of a State extends beyond its land territory and its internal waters, to a belt of sea adjacent to its coast, described as the territorial sea. . . .
> The sovereignty of a coastal State extends to the air space over the territorial sea as well as to its bed and subsoil.

Innocent passage is that which is "not prejudicial to the peace, good order or security of the coastal State" [Art. 14(4)]. *See also* Articles 15 to 17. On entry in distress, *see* H. Gary Knight, *The Law of the Sea: Cases, Documents and Readings* (Washington, D.C.: Nautilus Press, 1975) at 363-367.

5. Such zones have been established for customs regulation, neutrality, national security, environmental protection, fishing, and other purposes. *See* Knight, *supra* note 4, at 79-132.

6. The Convention on the High Seas [*done* April 29, 1958, 13 U.S.T. 2312 (1962), T.I.A.S. No. 5200, 450 U.N.T.S. 82, *in force* Sept. 30, 1962] provides in Article 1 that:

> The term "high seas" means all parts of the sea that are not included in the territorial sea or in the internal waters of a State.

Obviously, special limitations on freedoms of the high seas imposed by generally accepted special contiguous zones must be considered as modifying this definition.

7. Id., Art. 2.

8. *Supra* note 4.

9. Convention on the Continental Shelf [*done* April 29, 1958, 15 U.S.T. 471 (1964), T.I.A.S. No. 5578, 499 U.N.T.S. 311, *in force* June 10, 1964], Article 1.

10. Id., Art. 2.

11. Id., Art. 5(8). This provision requires ocean scientists to secure the consent of the coastal state about any research concerning the continental shelf

and undertaken there. However, although it also specifies that such state shall not normally withhold its consent if the request is from a qualified institution, representatives of the coastal state are permitted to participate, and the results are published.

12. Id., Arts. 4-5.

13. For background on the negotiations, *see* Knight, "Issues before the Third United Nations Conference on the Law of the Sea," 34 *La. L. Rev.* 155, 156-164 (1974); and Knight, "The Draft United Nations Convention on the International Seabed Area: Background, Description and Some Preliminary Thoughts," 8 *San Diego L. Rev.* 459, 477-486 (1971).

14. For accounts of the Caracas session, *see* Knight, "The Third United Nations Law of the Sea Conference: Caracas," *American Univ. Fieldstaff Reports*, vol. 18, no. 1 (October 1974); Stevenson and Oxman, "The Third United Nations Conference on the Law of the Sea: The 1974 Caracas Session," 69 *Am. J. Int'l L.* 1 (1975); and "Law of the Sea Briefing: Reflections on the Caracas Session on the United Nations Law of the Sea Conference," Occasional Paper No. 24, Law of the Sea Institute (December 1974). For accounts of the Geneva session, *see* "Hearing on Geneva Session of the Third U.N. Law of the Sea Conference" Before the Senate Commerce Committee, 94th Cong., 1st Sess. (June 3-4, 1975).

15. The delimitation of the baseline is itself a subject for negotiation in the conference, although little attention has been paid to the issue outside the archipelagic concept debates. Nonetheless, it should be understood that location of the baseline can have a significant effect on the seaward extent of any economic zone.

16. Informal Single Negotiating Text, UN Doc. A/CONF.62/WP.8, Parts I, II, and III (6-7 May 1975).

17. *See e.g.,* Statement of John Norton Moore Before the Subcommittee on Fisheries and Wildlife of the House Committee on Merchant Marine and Fisheries (May 19, 1975); Statement of Thomas A. Clingan, Jr., id.

18. Informal Single Negotiating Text, *supra* note 16, Part II, Art. 45.

19. G.A. Res. 2749 (XXV) (1970), 10 *Int'l Legal Materials* 220 (1971), adopted by 108 votes to none with 14 abstentions. In spite of the overwhelming support for Resolution 2749 in the voting, its value as evidence of customary international law is greatly reduced by the compromise nature and vague generality of most of its operative provisions.

20. Note 11 and accompanying text.

21. Informal Single Negotiating Text, *supra* note 16, Part III (Marine Scientific Research), Arts. 13-26.

22. Id., Part I, Art. 10.

23. Id., Part II, Art. 118.

24. Arctic Waters Pollution Prevention Act (C-202, 2d Sess., 28th Parl.,

18-19 Eliz. II, 1969-70). For an analysis of the legal-political framework of the act, *see* Bilder, "The Canadian Arctic Waters Pollution Prevention Act: New Stresses on the Law of the Sea," 69 *Mich. L. Rev.* 1 (1970).

25. *See, e.g.,* Bilder, "The Anglo-Icelandic Fisheries Dispute," 1973 *Wisc. L. Rev.* 37 (1973).

26. Deepwater Port Act of 1974 (Pub. L. 93-627; 88 Stat. 2126). *See also* the proposed regulations to govern activities at such locations in 40 *Fed. Reg.* 19,956 (May 7, 1975).

27. *See* Statement of John Norton Moore, "Hearings Before the Special Joint Subcommittee of the Senate Committees on Commerce, Interior and Insular Affairs, and Public Works," 93d Cong., 1st Sess. (Oct. 2, 1973); Statement of John Norton Moore, "Hearings Before the House Committee on Merchant Marine and Fisheries," 93d Cong., 1st Sess. (June 12, 1973).

28. Church v. Hubbart, 2 Cranch (6 U.S.) 187 (1804).

29. Philip C. Jessup, *The Law of Territorial Waters and Maritime Jurisdiction* (New York: George Jennings Co., 1927) at 95-96.

30. H.A. Smith, *The Law and Custom of the Sea* (London: Stevens, 1950, 2nd ed. 1959) at 29.

31. McDougal and Schlei, "The Hydrogen Bomb Tests in Perspective: Lawful Measures for Security," 64 *Yale L. J.* 648, 659-660 (1955). Reprinted by permission of The Yale Law Journal Company and Fred B. Rothman and Company.

32. Myres S. McDougal and William T. Burke, *The Public Order of the Oceans* (New Haven, Conn.: Yale University Press, 1962) fn. 125, at 48-49.

33. *Supra*, text accompanying note 26.

34. Statement of John Norton Moore, "Hearings Before the House Committee on Merchant Marine and Fisheries," 93d Cong., 1st Sess. (June 12, 1973).

35. *See* the embodiment of this principle in Art. 5 of the Convention on the Continental Shelf, *supra* note 9.

36. Pres. Proc. No. 2667, 3 C.F.R., 1943-1948 Comp., at 67 (1945); 13 *Dep't State Bull.* 485 (Sept. 31, 1945).

37. Toomer v. Witsell, 334 U.S. 385, 402 (1948).

38. *See, e.g.,* Francis T. Christy and Anthony Scott, *The Common Wealth in Ocean Fisheries* (Baltimore, Md.: Johns Hopkins University Press for Resources for the Future, 1965), *passim.*

39. Obviously, direct solar energy in the form of sunlight can be enjoyed by everyone without appropriation. We speak here of devices for the conversion of ocean-stored energy into electric or other energy forms. In that case it cannot be denied that appropriation is required for enjoyment.

40. *Supra* note 19.

41. "Notice of Discovery and Claim of Exclusive Mining Rights, and Request for Diplomatic Protection and Protection of Investment, by Deepsea

Ventures, Inc.," November 14, 1974 (filed in the form of a letter to Hon. Henry A. Kissinger, Secretary of State, U.S. Department of State).

42. Such bills have been introduced in the 92d, 93d, and 94th Congresses in the United States. The current version is S.713 (94th Cong., 1st Sess., February 18, 1975). *See also, Congressional Record*, February 18, 1975, p. S.1946 *et seq.*

43. Convention on the Continental Shelf, *supra* note 9, Art. 5(2).

44. Id., Art. 5(3).

45. Id., Art. 5(6).

46. Id., Art. 5(7).

47. For an analysis of the system of shipping safety fairways, *see* Knight, "Shipping Safety Fairways: Conflict Amelioration in the Gulf of Mexico," 1 *J. Mar. L. & Comm.* 1 (1969).

48. *See* Chapter 5.

49. *See* "Soviet Trawlers Frighten North Sea Gas-Rig Crew," *The New York Times*, February 28, 1975, p. 4, col. 5; "Soviet Trawlers Circle a Gas Rig Off England," *The New York Times*, February 27, 1975, p. 8, col. 2; "Gas Rig Is Besieged by Soviet Trawlers," *Washington Star*, February 27, 1975.

50. *See* "Officials, Government Eyeing North Sea Oil Rig Protection," *Baton Rouge Morning Advocate*, February 17, 1975; *Ocean Science News*, April 25, 1975, p. 1.

51. Outer Continental Shelf Lands Act, 67 Stat. 462, 42 U.S.C. §§ 1331-1343 (1953).

52. Id., § 1333(a)(1).

53. Id., § 1333(e)(1).

54. "European Agreement for the Prevention of Broadcasts Transmitted from Stations Outside National Territories," European Treaty Series, no. 53 (*signed* January 22, 1965; *in force*, October 19, 1967), 62 *Am. J. Int'l L.* 814 (1968).

55. "Treaty on the Prohibition of the Emplacement of Nuclear Weapons and Other Weapons of Mass Destruction on the Sea-Bed and the Ocean Floor and in the Subsoil Thereof," Arts. I and II, *done* February 11, 1971, 23 U.S.T. 701, T.I.A.S. no. 7337, *in force* May 18, 1972. The United States is a party to the treaty.

56. Id., Art. III(1).

57. "Contemporary Practice of the United States Relating to International Law," 69 *Am. J. Int'l L.* 666-667 (1975).

58. *See* Chapter 8.

59. Convention on the High Seas, *supra* note 6. Article 4 provides that "[e]very State, whether coastal or not, has the right to sail ships under its flag on the high seas."

60. Convention on the High Seas, *supra* note 6, Article 5(1).

61. *See, e.g.,* H. Meyers, *The Nationality of Ships* (The Hague: Nijhoff, 1967); and Boleslaw A. Boczek, *Flags of Convenience* (Cambridge: Harvard University Press, 1962). *See also*, Knight, *supra* note 4, at 393-403.

62. *See, e.g.,* Thamkrapalil Kochus Thommen, *Legal Status of Government Merchant Ships in International Law* (The Hague: Nijhoff, 1962); and Lillich, "The Geneva Conference on the Law of the Sea and the Immunity of Foreign State-Owned Commercial Vessels," 28 *Geo. Wash. L. Rev.* 408 (1960).

63. Convention on the High Seas, *supra* note 6, Article 26(2) and (3).

4 International Political Implications of Ocean Thermal Energy Conversion Systems

Ann L. Hollick

Introduction

The international political implications of the development of a capability to recover energy from thermal differences in the oceans are twofold. The first is the effect that significant numbers of OTEC plants will have on the conduct of international relations. The second is the reverse effect, namely, the impact of the prevailing international political system on the use of OTEC plants in the world ocean. The latter relationship would appear to be more significant than the former.

As a new marine technology, the role that OTEC will play is similar to that of many other scientific and technological advances which are facilitating intensive use of the oceans. The consequence of such technologies is the territorialization of what were formerly nonnational areas, open to free use by all. Apart from strengthening the relative power of those states which already possess ample coastal areas, new ocean uses will *not* bring about a fundamental change in the international political system comparable to the discovery of nuclear energy. Like other technological developments that have promoted new uses of the oceans, the effect of OTEC on international politics will be that of a new complicating factor. OTEC plants may generate political controversy in terms of the activities they support or the products they produce. However, apart from the possibility that OTEC devices, in sufficient numbers, might alter the thermal layers of the oceans, there is nothing intrinsic to the operation of the OTEC plant itself that is likely to generate political problems. Indeed, from an environmental point of view, ocean thermal energy is regarded by its proponents as a relatively benign energy source.

The second and more significant causal relationship is the effect that international politics will have on OTEC. International political factors of particular importance to OTEC include political geography as well as the structure and operation of the international political system. Political geography is particularly important to the deployment of OTEC plants because of the implications of processes presently underway in the Third United Nations Conference on the Law of the Sea (UNCLOS III). These processes point to substantial changes in the legal regime of the oceans in the near future. The structure of the international political system is relevant to OTEC insofar as it will set the broad constraints on political interactions in the oceans. Operational considerations affecting the use of ocean thermal energy include the nature of the political actors and the international rules which govern their interactions relating to the siting and use of plants.

The examination of the relationship between OTEC and international political questions is rendered difficult at best by the time frame in which we are operating. Substantial numbers of OTEC plants are not expected until the twenty-first century. One projection is for the operation of a full-scale 100-MW floating demonstration power plant by 1988 with commercialization to begin by 1990 in the form of power plants having capacities ranging from 100 to 1000 MW each. Not until the year 2000, according to this speculation, are there likely to be over 100 power plants producing in the range of 10 to 60 GW.[1] For purposes of the following discussion, it is assumed that the international political problems that will surround the use of OTEC will mount as the number of power plants increases and as they are located at ever greater distances from the host state. At the earliest, moreover, we are considering the interaction between OTEC and an international political situation that will exist 25 years in the future. The difficulties of such "futurology" are apparent and at the outset merit a disclaimer on the speculative nature of such an enterprise.

Political Geography of OTEC Operations

As noted above, the impact of the widespread use of ocean thermal energy on international politics is likely to be that of increased complexity rather than that of a fundamental transformation of international political relations. In the area of political geography, however, OTEC plays a causal role as one of a number of new technological developments leading to new activities in the oceans. These new activities have stimulated the interest of coastal states in extending their control over vast areas of ocean space. The consequences of these extensions will be substantial for the future use of OTEC.

The question of the geographical location of OTEC plants and related operations has two aspects. The first and broadest is the determination of the areas where climatic conditions make OTEC operation feasible and economically attractive. The second aspect engages the complex of questions with jurisdictional implications, such as which types of national, international, and mixed jurisdictions are exercised by which authorities and according to which rules and regulations.

General Area of Operation

Fundamental to the operation of an OTEC plant is the existence of substantial differences in temperature between surface and subsurface ocean waters. The greater the difference, the more efficient the process. Present design concepts suggest the need for temperature differentials (ΔT°) of at least 40°F for economic operation of an OTEC plant. A related variable is the distance between the surface and subsurface layers of differing temperatures. The greater the distance, the longer the cold- and warm-water intake pipes must be and the greater the resulting expense.

On a global basis, optimal temperature and water-depth conditions are found within a belt around the equator. In these regions, comprising 20 percent of ocean space, during the winter surface waters reach an average of 77°F while subsurface layers within 1500 ft or less remain at temperatures of 40°F and below.[2] Conservative estimates place the OTEC-suited areas between 10°S and 10°N latitude. Suitable areas, however, may be found up to 20°N and 20°S latitude and beyond, depending upon ocean currents and evolving technology.

It is interesting to note that, with the exception of the United States, the countries situated within this equatorial band are not major consumers of energy at the present time (see Figure 4-1). A number of these countries do have large populations, however, and their energy consumption levels may change substantially by the time OTEC is widely available. Countries adjacent to areas suitable for OTEC include Nigeria, the Ivory Coast, Guinea, Senegal, Kenya, Somalia, India, Burma, Thailand, Indonesia, the Philippines, Australia, Fiji, Ecuador, Colombia, Venezuela, Mexico, Brazil, the Central American and Caribbean states, and the United States. Some of these countries are endowed with abundant raw materials (rubber, wood, bauxite, iron, steel) and might welcome a new source of energy to facilitate the development of local industry. Others, of course, already enjoy alternative forms of energy (petroleum and hydroelectric) as well as resources suitable to local processing. In general, however, OTEC offers the countries of the equatorial regions not only the opportunity to generate electricity in their offshore waters for onshore use, but also the possibility of transferring energy-intensive or environmentally harmful production activities to offshore areas. The use of OTEC by these countries as well as nations in other regions of the world raises the issue of the types of legal jurisdiction that may affect the deployment of OTEC devices in the world oceans.

Legal Jurisdiction: The Law of the Sea

To speculate with a high degree of accuracy on legal regimes as well as political situations as far off in the future as 25 years is difficult. The only certainty is that the present legal regime for the oceans will not persist. Indeed, if trends apparent within the Law of the Sea Conference are any indication, the present legal regime may not last beyond 1976.

The Law of the Sea Conference has been meeting annually since 1973, and preparatory work began even earlier. The only product of its meetings to date is

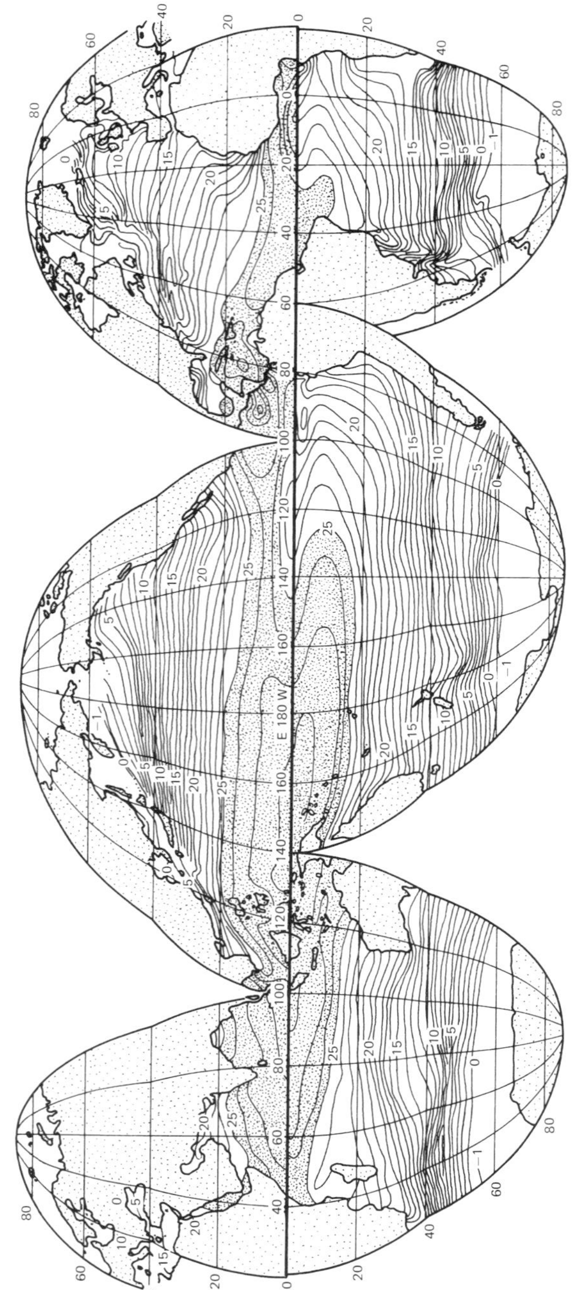

Source: Lockheed Missiles & Space Co., Inc., Ocean Systems

Note: Regions with winter average of 25°C (77°F) account for 20 percent of ocean surface.

Figure 4-1. Regions of Utilization.

a document referred to as the Informal Single Negotiating Text (ISNT) produced by the chairmen of the three main committees of the conference in May 1975.[3] Although this text is not a negotiated document and is not binding on the delegations, it is expected to serve as the basis for future negotiations. For present purposes, the ISNT is useful in illustrating the possible future regimes that may pertain in the oceans. Even if a treaty based on the ISNT were not adopted by the conference, many of its provisions on coastal state jurisdiction would be adopted unilaterally. In effect, the ISNT embodies the expectations of many nations and the behavioral norms that have been generated in the course of UNCLOS III.

Perhaps the salient characteristic of the ISNT is that it creates a future legal regime for the oceans that will be far more complex than that established by the 1958 conventions on the law of the sea. Equally notable is the fact that the ISNT posits a dramatic increase in the area of ocean space under national jurisdiction. Depending upon the construction of baselines and upon agreement on outer limits to the continental margins, the ISNT provisions ensure that over 40 percent of ocean space will fall under national jurisdiction of one form or another.[4] Indeed, it has been suggested that coastal state extensions underway in UNCLOS III are simply the first steps toward an ultimate division of the oceans. Indications of this long-range outcome may be found in claims of some nations to living resources "adjacent to" their 200-mi coastal state resource zone.[5] Such claims may be supplemented by claims to mineral resources beyond the continental margin, particularly if the present conference does not agree on a regime for deep seabed mining. For the next few years, however, the coastal state extensions provided for in the ISNT should occupy coastal state energies—if not in effectively managing the areas, at least in attempting to resolve the numerous jurisdictional conflicts that are bound to ensue.

The types of national jurisdiction will grow in number as well as increase in size. First the baselines from which other zones are measured will be lengthened to enclose larger areas as internal waters. In the case of islands and archipelagoes, the straight baselines will enclose archipelagic waters. They will be drawn to link the outermost points of the outermost islands and drying reefs of the archipelago, thereby enclosing within national jurisdiction vast areas of ocean space where OTEC devices can operate economically (Indonesia, Philippines, Fiji, Mauritius, etc.).

Beyond the archipelagic and coastal baselines, the ISNT provides for the establishment of various forms of national jurisdiction. These include territorial seas up to 12 mi in width, contiguous zones up to 24 mi wide, an exclusive economic zone up to 200 mi, and a continental shelf extending beyond 200 mi to the outer edge of the continental margin.[6] Within each of these areas the coastal state will enjoy a different set of rights. Some of these rights more clearly extend to the control of OTEC plants than others. The only specific reference, however, to energy-producing devices is found in Part II of the ISNT, which deals with the exclusive economic zone. It provides the coastal state with "exclusive jurisdiction with regard to (i) other activities for the economic exploitation and exploration of the zone, such as the production of energy from the water, currents and winds; and (ii) scientific research."[7] Article 48 strengthens coastal state rights that would apply to the attempted use of OTEC by foreign countries in the 200-mi zone. It provides that the coastal state shall have the "exclusive right to construct and to authorize and regulate the construction, operation and use of (a) artificial islands; (b) installations and structures" used for economic purposes or which may interfere with the rights of the coastal state in its zone.

The use of OTEC within 200 mi of another state for activities other than the transmission of energy to shore would, according to this proposed legal regime, be very problematic. Presumably, OTEC devices used within the economic zone for economic activities and scientific research or falling within the definition of an artificial island or installation will be subject to the exclusive jurisdiction of the coastal state. Less clear is the status of OTEC devices that are not used to support economic activities or scientific research in the zone. Also uncertain is exactly what constitutes an artificial island or installation.

Beyond the 200-mi exclusive economic zone, the ISNT provides the coastal state with "sovereign rights" for the purposes of exploring and exploiting the natural resources of the continental shelf.[8] The continental shelf is used in the ISNT to include the entire continental margin (see Figure 4-2). Unfortunately, the edge of the continental margin is geologically difficult to determine and has been the subject of much discussion in UNCLOS III. In the case of Australia, off whose shores ocean thermal energy conversion would be feasible, continental shelf jurisdiction may be extended beyond 400 mi from shore depending upon the definition of the margin ultimately adopted. The ISNT is unclear on whether

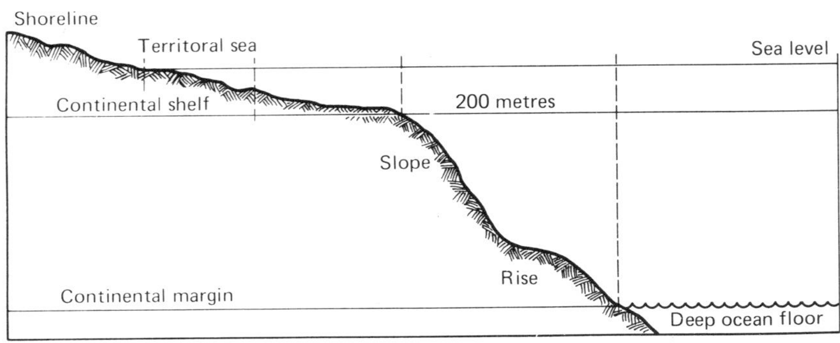

Figure 4-2. Idealized Topographic Profile of Continental Margin (Vertical Scale Exaggerated).

coastal state consent would be required for OTEC plants operating in waters above the legal continental shelf. Certainly a plant used to support "research concerning the continental shelf and undertaken there" would require coastal state consent. Similarly, an OTEC plant supporting activities relating to the exploration and exploitation of the shelf resources would seem to require the "express consent" of the coastal state. Less clear is whether an OTEC plant engaged in other activities might fall under the consent requirement simply because it is over the shelf or because it is moored to the shelf.[9]

A similar complex of questions is raised by the provisions of the ISNT dealing with the deep seabed beyond the limits of national jurisdiction. The Committee I text provides for comprehensive powers to be vested in an international seabed authority. These include the "regulation and supervision by the Authority" (Part I, Article 6) of "all activities of exploration of the Area and of the exploitation of its resources, as well as other associated activities in the Area including scientific research" (Part I, Article 1). The term *associated activities in the Area* is critical here. It may be broadened to cover OTEC devices supporting scientific research which is related only indirectly to the seabed and its resources. Efforts may also be expected to extend the concept of "associated activities" to OTEC plants supporting the processing and refining of manganese nodules on ocean platforms. And, of course, a bottom-moored OTEC ultimately may be deemed to come under the regulation of the seabed authority by virtue of its physical attachment and regardless of the OTEC activity.

How national and international regulatory regimes affecting future OTEC usage will evolve is uncertain. Developed countries are strongly opposed to ISNT provisions for thoroughgoing authority controls in the deep seabed. Depending upon the willingness of the less developed nations to compromise, the terms for a deep seabed regime ultimately may be modified to a less restrictive regime. In comparison, the political forces opposing extended coastal state jurisdiction are relatively weak. The likelihood, therefore, is that ISNT provisions placing substantial portions of the oceans under national jurisdiction will be internationally accepted. The question, of course, remains as to whether these national areas will be extended to a full-scale coastal state partitioning of the oceans by the year 2000, when a large number of OTECs will come into operation. To some extent this will depend upon whether a viable seabed mining regime is internationally agreed upon. Certainly the division of the oceans among nation states according to median lines would have a substantial effect on the problems of OTEC deployment. For purposes of the following discussion, however, the more conservative assumption is made that the oceans will be divided into areas of national as well as nonnational jurisdiction.

International Political Factors Affecting OTEC

Among the political variables that will influence the utilization of OTEC plants are the structure of the international system, the major actors that will have a role in OTEC use, and the prevailing modes of political interaction among the actors. The international political configuration of the twenty-first century, when large numbers of OTEC devices could be available, is highly uncertain. Just as if would have been difficult in 1950 to correctly predict the prevailing political conditions of today, so too the year 2000 is a dubious target for speculation.

To be sure, the antecedents of the present are found in the past. In a similar fashion, the configuration of today may yield some intimations of the future. The dominant structure of the present political scene includes a devolving East-West tension and a relatively recent North-South conflict over the distribution of global wealth. Within this context, the salient international political actors are nation-states. Other actors include subnational players such as terrorist organizations and multinational corporations as well as supranational international organizations. The interaction of these actors is characterized by conflict as well as cooperation, reflecting a substantial degree of independence as well as some perception of interdependence.

International Political Actors

The international political actors that will be directly or indirectly involved with OTEC activities in the year 2000 may include national, subnational, or supranational entities. A recurrent debate in the study of international relations concerns the long-term viability of the nation-state as the guarantor of individual and community values and consequently its viability as the central actor in international politics. One side of the debate points out that the functions previously performed by the nation-state are increasingly carried out by transnational organizations such as multinational corporations or international agencies. This trend, according to this point of view, is inevitable since it is based on economic, environmental, social, and political factors that increasingly dictate interdependence. Under this scenario, the nation-state will eventually be replaced by ever larger, increasingly integrated institutions such as the European Community which will more adequately serve human needs and interests.

The counterview in this debate is that the growth of activities by transnational organizations does not spell the demise of the nation-state. Indeed, it reflects the continued viability of the nation-state as the central actor in the political system by using other nonnational entities to maintain that position. This school of thought argues that the nation-state is the instrument through which societies will continue to fulfill their needs even in those situations where international interdependence exists. Total global interdependence is seen as neither preordained nor self-evident.

Certainly trends exist today that would seem to buttress both sides of the argument. Strongly nationalist forces are active in international forums at the same time that there are growing efforts to resolve pressing economic and other problems at the international and transnational levels. These contradictory tendencies do not resolve the debate or indicate the international actors that will affect OTEC activities in the year 2000. In this situation, it is perhaps wiser to extrapolate from the present and assume that nation-states will retain a major role in 25 years and that nonnational actors such as terrorist organizations, multinational corporations, and international organizations will be among those having an effect on the deployment and use of OTEC devices.

Structure of the International System

Nation-states operate within a framework structured by cooperative alignments as well as by political conflict. The relations among nations, and hence the structure of the system, change as the relative power base of states alters and as salient issues or ideologies come and go. The structure of the present international system is characterized by two overlapping lines of conflict—North versus South and East versus West. Whether these divisions will endure to the year 2000 and affect OTEC-based activities is uncertain.

The East-West division may be expected to change as the preeminence of the two superpowers is reduced by the growth of other centers of economic, military, and political power. Similarly the ideological aspect of the confrontation is declining and may continue to decline with increased exchange between East and West and the absence of an audience—namely, the developing world—to be won over.

The development of the North-South confrontation has played a substantial role in dampening East-West tensions. The South no longer accepts the role of the passive recipient of aid to be competitively dispensed by states seeking ideological victories. However, in the recent successes of the OPEC countries vis-à-vis the industrial nations lie the seeds of a future diminution of Southern solidarity and hence of the sharply focused North-South conflict. Among the power constellations that may emerge are a series of producer-consumer differences leading ultimately to a delineation between big powers, middle powers, and the poorer nations, rather than between rich and poor alone.

Were a North-South conflict to persist in present or modified form, it would pose certain problems for the deployment of OTEC-supported activities. These problems stem from the obvious geographical and economic facts of the situation. Areas suitable for generating energy from thermal differences in the oceans are, for the most part, in the equatorial regions, while the major energy-consuming states are, at least for the present, found in the temperate Northern areas. Yet it is the developed countries that have the capital needed to construct OTEC plants. Once the capital is available, however, the technology of OTEC is such that plants can be operated by nationals of even the least developed countries.

Developed-developing country relations regarding the use of OTEC for economic or scientific purposes may be expected to range between two extremes. At the cooperative end of the spectrum, the developed countries may offer OTEC technology, training, and financing in return for energy supplies or other commodities. At the other extreme, developing states may deny access for OTEC-based activities to ocean areas in a given region. The new "smart" missiles make an attack by the coastal state on unauthorized OTEC installations relatively easy. In any event, it seems unlikely that developed countries would conduct OTEC-based economic activities where there was a high risk of conflict—i.e., without permission in the economic zone of another state. Production activities in the area beyond 200 mi would be more problematic due to regional sensibilities and potential jurisdictional claims by littoral states.

The use of OTEC installations for military activities raises a similar set of issues with even fewer international guidelines. Partial demilitarization of the seabed is provided for by the Treaty on the Prohibition of the Emplacement of Nuclear Weapons and Other Weapons of Mass Destruction on the Seabed and the Ocean Floor and in the Subsoil Thereof.[10] Commended by the 25th Session of the UN General Assembly in December 1970, the treaty came into force in May 1972. It prohibits parties to the treaty from emplacing or emplanting on the ocean floor beyond 12 mi any nuclear or other types of weapons of mass destruction, or structures, launching installations, and other facilities designed for storing, testing, or using such weapons. States are free to emplace such devices within 12 mi of shore. The seabed implantation of weapons that do not fall into the category of being capable of "mass destruction" is not prohibited and presumably might be supported by a fixed OTEC device. Dynamically positioned OTEC devices could, of course, offer a means around the prohibition since, even if occasionally moored, they would not be considered emplaced or emplanted. The ISNT offers no prohibition on the use of military activities in the economic zone of another state, as long as those activities do not fall under the category of scientific research, economic exploitation, or artificial islands or installations. Hence, this omission might be taken as an indication that military activities are contemplated in these areas, presumably insofar as they are mobile and do not fall under the prohibitions of the Seabed Demilitarization Treaty. Even if the Law of the Sea Conference adopts a treaty without restrictions on foreign military activities in the economic zone, the objections of the littoral states to military activities both within and beyond the 200-mi zone are easy to foresee. Efforts to ban military activities carried out by nonlittoral states in the Indian Ocean indicate developing-country preferences for excluding superpower rivalries in favor of strictly regional confrontations and competition.

Clearly the East-West systemic conflict is linked to the North-South division in a number of ways. Like that of the North and South, it may range from cooperation to conflict along a number of functional areas—scientific, economic, and military. OTEC may be used to further cooperative scientific research at the one extreme. At the other extreme, the superpowers may use OTEC to support sonar arrays or even offensive nuclear weapons (if the OTEC plant is mobile). Both the U.S. and U.S.S.R. signed the Seabed Demilitarization Treaty and would presumably not emplant such offensive weapons on the deep seabed. In between these extremes, U.S. and U.S.S.R. economic competition may be carried to OTEC-based activities. Fish farming, deep-sea mining of fuel, and water ports in support of commercial shipping are among the possible uses of energy generated by ocean thermal energy plants. In general OTEC will provide new means to continue preexisting relations between East and West. Like other new uses of the oceans, OTEC will complicate rather than change the interaction. This is true of the military area in particular. Present deterrent systems rest on the retaliatory capability of submarine forces operating in an opaque environment. While OTEC offers an opportunity to scatter sonar arrays around the oceans, it also creates more difficulties for antisubmarine warfare (ASW). The more objects in the ocean near which a submarine can hide, the more complicated the task of sonar detection.[11]

In conclusion, the future of East-West relations, like those between the North and South, depends very much on how they are handled today. Where zero-sum approaches are adopted by one or both sides to a rivalry, the tension will clearly persist. If, however, a positive-sum interaction is set in motion, the benefits to both sides may well exceed those expected from continuing confrontation.

Operational Issues Affecting OTEC

While the overall structure of the system will set the broad constraints on the development of OTEC, the direct operational problems will stem from the local interactions among nation-states and nonnational actors. These will be responsive to prevailing regulatory systems and norms of international behavior. Problems for OTEC utilization may be grouped for purposes of discussion into two types—those occurring when OTEC plants are located within areas of national economic jurisdiction, and those relating to OTEC use beyond. The problems pertinent to ocean thermal energy conversion within areas of national jurisdiction stem from the intensive use made of near-shore areas as well as from the inevitable jurisdictional disputes. The problems that pertain to OTEC use beyond national areas will be limited largely to those of a jurisdictional character.

OTEC in Areas of National Jurisdiction

The extension of coastal state exclusive economic jurisdiction to 200 mi will create a coastal state management responsibility for the most heavily used portions of the oceans. Recreational uses, fishing, petroleum recovery, shipping, and scientific research are among the activities which the coastal state will have to coordinate. In addition to these new responsibilities, extensions of coastal state jurisdiction will increase the number of national neighbors, and hence the number of border problems, particularly for states fronting on semienclosed seas. For others, borders with current neighbors will be lengthened. In addition to the problems of congestion, therefore, there will be difficulties of a territorial nature, ranging from boundary disputes to dissatisfaction with the activities of an adjacent state in its own zone.

In this context, the advent of ocean thermal energy will simply further complicate an already complex situation. The least controversial scenario for OTEC utilization is that of a coastal state producing energy from its zone for transmission to shore. This becomes more problematic where the numbers of OTEC plants deployed are substantial enough to alter the thermal layers within the zones of adjacent states. Of course, the interaction becomes progressively more difficult where OTEC is used for offshore industry or other activities.

The types of activities which ocean thermal energy may support are numerous. Given the concentration of populations along coastal areas, production activities offshore offer easy access to centers of consumption. In addition to providing support for energy-intensive industrial processing, ocean thermal energy might be converted to other forms of energy (i.e., hydrogen) and shipped to shore. Of course, if populations move offshore for production activities, they may well do so in order to live. In the creation of offshore islands and habitats, OTEC will play a major support role. The environmental impact of such offshore centers of human habitation will certainly be of concern to neighboring states. Moreover, it is not difficult to envision a situation in which a coastal state encourages the movement offshore of production activities which are physically dangerous or environmentally damaging. Supported by OTEC, these activities may be moved to the outermost reaches of a state's economic zone—with consequences of concern to neighboring states.

The variations of these scenarios are numerous. For instance, some states might be willing to use their offshore areas as "OTEC havens." In exchange for appropriate fees or taxes, they might open their zones to activities of states or groups outside the region. These could include military surveillance of or threat to other states in the region, environmentally damaging production activities, or simply multinationals seeking to avoid financial or environmental restrictions. The activities in such OTEC havens would obviously concern neighboring states.

A different situation is conceivable: terrorists based in one state might disrupt offshore activities in the zone of its host state or of adjacent states. The inability of some countries to control paramilitary domestic groups can have harmful consequences for neighboring countries. While the vulnerability of an offshore OTEC plant would seem to make it an ideal target for terrorist activities, this is likely to occur only when populations move to offshore centers.

Terrorists seek visibility for their actions, and that is more readily available in heavily populated areas than at a remote OTEC installation.[12]

The use of OTEC devices for national military purposes is far more likely than paramilitary or terrorist activity. Since military uses are not included within the economic competence of the coastal state in the present law-of-the-sea discussions, military activities in the economic zone may occur with or without the concurrence of the coastal state, much less that of neighboring coastal states. In using its own zone for military purposes, the coastal state may find that a roving or dynamically positioned OTEC serves as a useful base for conventional or nuclear weapons or for intelligence gathering and surveillance against nearby or distant states. An OTEC device would appear to offer the coastal state some of the advantages of a submarine over hardened missile sites in terms of mobility. Moreover, when they are deployed in larger numbers, it may prove difficult to determine the purpose of all OTEC plants. Implicit in the absence of restrictions on military activities in the zones of other states at the Law of the Sea Conference is the preference of the major powers to retain the high-seas character of the waters beyond 12 mi. As the coastal state's economic control of the 200-mi zone is consolidated, however, there will be increasing objection to unauthorized military activities and a move to establish a regime of complete national sovereignty for the 200-mi zone.

OTEC beyond National Jurisdiction

The problem of OTEC utilization beyond 200 mi is less one of reconciling multiple uses than one of jurisdictional uncertainty. This stems from the ambiguity about continental shelf rights, the uncertain rights of a seabed authority, and the likelihood that coastal states will want to extend their jurisdiction beyond 200 mi in the not-too-distant future.

OTEC plants used on or above the continental margins beyond 200 mi may be subjected ultimately to coastal state control. The history of the Convention on the Continental Shelf has been one granting to the coastal state ever-growing powers in the area. Brazil has used its continental shelf rights to ban scientific research conducted over the entire margin, and the United States has strictly regulated artificial installations over and attached to its shelf. Canada has used the shelf rights in a different fashion to claim fisheries in the superjacent waters. The possibilities are numerous, and history suggests that all available legal arguments will be employed to control OTEC activities over the margin beyond 200 mi.

The problems that OTEC users will face beyond 200 mi and the margin include potential difficulties with an international seabed authority or with coastal states objecting to activities conducted adjacent to their respective economic zones. If a strong international authority is agreed to by the Law of the Sea Conference, it will certainly attempt to regulate OTEC installations used in support of scientific research and processing of manganese nodules. Alternatively, states of a region, such as the Indian Ocean, may band together to protest the use of OTEC devices to support military activities of outside states. Whether this protest would be advanced to prohibit economic activities is less clear.

An additional form of jurisdictional uncertainty may arise if OTEC plants come to be used under a flag-of-convenience system. Multinationals wishing to move beyond national jurisdiction for tax or environmental reasons may find states willing to grant them their flags in exchange for the appropriate registry fees. Whether the OTEC device is moored, is dynamically positioned, or roams freely, it could claim the nationality of the registry state regardless of ownership, crew, or activities. This suggests that many states could become involved in OTEC operations. Such diffusion of ownership would lead to reciprocal vulnerability to the effects of ocean uses and hopefully to greater cooperative efforts.

This raises the issue of international regulation in its most acute form. Would such regulation be based on the fact that the OTEC device moves freely, on the fact that it is involved in the recovery of energy from the oceans, or on the character of the activities it supports? Different international organizations would be called for in each case. If its qualities as a semimobile structure were considered preeminent, then the Intergovernmental Maritime Consultative Organization would be the appropriate international regulatory organization. If, however, its energy-generating activities were considered of paramount importance, then it might fall under the purview of an expanded International Energy Agency.[13] Finally, if the activities with which the OTEC device was associated were deemed most significant, then a variety of international organizations might exercise some regulatory control. An OTEC device used for fish farming could be the object of the Food and Agriculture Organization's attention. OTEC plants supporting scientific activities might fall under the aegis of the International Oceanographic Commission. The World Meteorological Organization might have coordinating responsibility for OTEC plants supporting satellite and weather research. A seabed authority would want to regulate OTEC mineral-related activities. Even military activities might be undertaken under the auspices of regional alliances such as SEATO or NATO.

Since the nature of the OTEC-based product or activity will, in most cases, be the subject of legitimate international concern, it is likely that regulatory jurisdiction will be exerted by some or all of the above-mentioned organizations. The resulting mosaic of regulatory bodies is likely to be quite complex in the absence of a single all-powerful ocean organization.

If we look into the future, we can foresee a regime for the oceans which includes the following components: (1) general legal principles which are arrived at via customary law or international negotiation and which are agreed to in treaties, declarations, or resolutions; (2) a variety of regional and subregional arrangements or agreements; (3) a mosaic of international regulatory authorities concerned with different aspects of the oceans; and (4) reverberations from the prevailing conflicts and alignments in other areas of the international political and economic system. The salient feature of this regime will be its fluidity or uncertainty. The process of exercising new forms of jurisdiction in the oceans will be characterized by trial and error as some arrangements are deemed inadequate and replaced by others. An integral part of this process of testing will be continued conflicts ranging from serious border disputes to minor differences over scientific management questions. In this context, OTEC-supported activities will offer another source of conflict as well as an opportunity for cooperative arrangements.

Notes

1. U.S. Energy Research and Development Administration, *A National Plan for Energy Research, Development and Demonstration: Creating Energy Choices for the Future*, vol. 2 (June 1975), (ERDA-48). The projection made by Lockheed Ocean Systems is for the production of four plants by 1990 and of 150 plants by the year 2000, if the price of energy remains high and the recommended work schedule is maintained. By way of comparison, in 1973 the United States had a total installed capacity of 390 GW.

2. *See* Chapter 1.

3. Informal Single Negotiating Text, UN Doc. A/CONF.62/WP.8, Parts I, II (7 May 1975), III (6 May 1975).

4. Of the 20 percent of the oceans where OTEC will operate most effectively, over 40 percent would constitute national areas.

5. Draft Articles on Fisheries in National and International Zones in Ocean Space: Submitted by the Delegations of Ecuador, Panama, and Peru, UN Doc. A/AC.138/SC.II/L.54 (10 August 1973), Article I.

6. Informal Single Negotiating Text, *supra* note 3, Part II.

7. Informal Single Negotiating Text, *supra* note 3, Art. 45. The origins of this article are somewhat obscured by the off-the-record, informal negotiating procedures of UNCLOS III. The article first appeared in an informal document called the "Evensen Text" which was submitted to the chairman of Committee II. Within the Evensen negotiating group, the inclusion of reference to energy-producing devices was proposed by several Latin American states. The reaction of the developed states was neutral. They reportedly viewed such a provision as duplicating Article 48, which already accorded the coastal state jurisdiction over all economic installations and structures in the economic zone.

8. The geological continental shelf comprises that gradually sloping portion of the continental margin extending from the shoreline to the more sharply

inclined continental slope. The abrupt transition from the shelf to the slope occurs around the world at various depths and distances from shore. The legal definition of the continental shelf has been flexibly altered as the interest in offshore resources has progressed. In the 1958 Convention on the Continental Shelf, it was defined as extending to a depth of 200 m or beyond that to a depth admitting of exploitation. The "exploitability clause" has proved highly adaptable to the perceived interests of coastal states.

9. The distinction between moored and dynamically positioned OTEC devices has economic as well as potential legal implications. The net energy output of a mobile OTEC device is necessarily lowered by the diversion of energy for positioning purposes.

10. Treaty on the Prohibition of the Emplacement of Nuclear Weapons and Other Weapons of Mass Destruction on the Seabed and the Ocean Floor and in the Subsoil Thereof, February 11, 1971, 23 U.S.T. 701, T.I.A.S. no. 7337 (*in force* May 18, 1972).

11. For an excellent and detailed discussion of current and projected military uses of the oceans, *see SIPRI Yearbook of World Armaments and Disarmament.*

12. For an excellent discussion of the problems of terrorism facing offshore installations in the North Sea, *see* Johan J. Holst, "The Strategic and Security Requirements of North Sea Oil and Gas," in Martin Salter and Ian Smart, eds., *The Political Implications of North Sea Oil and Gas*, IPC Science and Technology Press, 1975.

13. IEA's major function to date has been as a coordinating mechanism for developed oil-consuming countries.

5 Spatial and Emerging Use Conflicts of Ocean Space

Byron Washom

Ocean space—the surface of the seas, the water column, the seabed and its subsoil—is, however, by far the largest and most valuable region of our planet which still awaits full utilization by man.

Arvid Pardo[1]

Introduction

The emergence of ocean thermal energy conversion represents but one of many possible new uses of the ocean in the near future. New uses, when combined with the intensification of existing ones, will undeniably lead to conflicts over use of ocean space. As new technologies such as OTEC are developed, they should be assessed as to their role and impact in the development and wise use of the ocean's resources. Presently, OTEC serves as a futuristic and potentially replenishable source of energy for on-land energy consumption, fuel production, or *in situ* processing. The questions examined in this chapter, therefore, concern the possible ocean space conflicts between OTEC and a variety of future ocean uses and how extensive these conflicts are likely to be.

It is assumed that OTEC plants would be located within three possible regions: (1) less than 3 mi from a coastal state; (2) between 3 mi and the extent of a coastal state's jurisdiction; or (3) beyond a coastal state's jurisdiction. The moored structure will have substantial area exposed to the surface, within 50 ft of the surface, or readily surfaceable during emergency situations. Finally, the cold-water pipe will extend a minimum of 2000 ft below sea level and as much as 4000 ft.

To assess the spatial and use conflicts of OTEC, this chapter examines the nature of future activities and how OTEC might conflict with such activities. Identifiable conflicts can be resolved through technology development, national laws and regulations, international treaties and agreements, or dispute settlement. Collectively, these actions would minimize, if not eliminate, the spatial and use conflicts of OTEC in the ocean space.

Present and Emerging Uses of Ocean Space

What is a likely scenario of ocean utilization in 1985 or 2000? Current situations and trends provide valid indicators concerning the volume and efficiency of ocean resource utilization. Although one should be cautious in reiterating the overoptimistic claim that scarcity of land-based resources will instigate pursuit of oceanic alternatives, the fact remains that recent events have seriously threatened states whose land-based supply of essential products is alternately available from the oceans. With the realization of advancements in technology and the impact of resource cartels, most countries are asserting rights to share in the potential benefits of the sea. Should a 200-mi exclusive economic zone be adopted, it would place 37 percent of the world's ocean space under the partial control of the individual coastal states.

Three factors will accelerate the rate of ocean utilization over the next 25 years: economic necessity, jurisdictional permissibility, and available technology.

In economic terms, Table 5-1 presents a likely scenario of ocean use in the years 1985 and 2000. These figures are elaborated in the following two sections. This analysis is purposefully divided into two divisions of extractive and nonextractive uses to illustrate the nature and impact of each.

Extractive Uses

Extraction of ocean resources can be of those that are either regenerative or nonregenerative. The greatest activity concerning regenerative resources deals with fishing, where tremendous technological advancements have increased the efficiency of locating, catching, and processing the fish. Light amplifiers, low-level television, the Earth Resources Technology Satellite (ERTS), and weather satellites have all been innovations of the last 5 to 7 years. Collectively they have reduced the searching time and increased the probability of locating stocks to a level rapidly approaching maximum sustainable worldwide yield. Fishing vessels are now capable of catching and transporting 500 tons of raw fish. Although the U.S. fleets have leveled off at approximately 2.4 billion lb of landings, imports have risen to 1.8 billion lb/year.[2] The total economic value of U.S. fish consumption was $1.89 billion in 1972. The processing of caught fish has expanded to optimal utilization of all parts of a fish to the point where very little processing waste is now discharged. Other regenerative resources include shellfish for their lime, algae for their algin, and drugs from the sea.

Nonregenerative extractive resources are those located on the seabed and subsoil. Foremost in value of such resources are petroleum and natural gas, which will undergo a tremendous increase in activity in the next 15 years. More than 85 countries are engaged in offshore activities, of which 32 are actually producing some 16 percent of the world's oil and 6 percent of the world's natural gas. This production is expected to quadruple by 1980.[3] Technology is keeping pace with the world demand for petroleum by providing the capability of continuously operating at greater depths and in adverse environments. Other subsurface deposits include sulfur, coal, salt, and potash.

The seabed contains a variety of mineral resources ranging from beach sand

Table 5-1
Estimated and Projected Primary Economic Value of Selected Ocean Resources to the United States, By Type of Activity, 1972/73-2000, In Terms of Gross Ocean-Related Outputs
(In billions of 1973 dollars)

Activity	*1972*	*1973*	*1985*	*2000*
Mineral resources:				
Petroleum		2.40	9.60	10.50
Natural gas		0.80	5.80	8.30
Manganese nodules			0.13	0.28
Sulfur		0.04	0.04	0.04
Fresh water		0.01	0.02	0.04
Construction materials		0.01	0.01	0.03
Magnesium		0.14	0.21	0.31
Other			0.01	0.02
Total		3.40	15.82	19.52
Living resources:				
Food fish	0.74		0.95- 1.58	1.37- 4.01
Industrial fish	0.05		0.05- 0.08	0.05- 0.14
Botanical resources	*		*	*
Total	0.79		1.00- 1.66	1.42- 4.15
Nonextractive uses:				
Energy			0.58- 0.81	3.78- 6.03
Recreation	0.70-0.97		1.12- 1.50	1.64- 2.53
Transportation	2.57		4.40- 6.21	6.88-11.41
Communication	0.13		0.26- 0.36	0.44- 0.85
Receptacle for waste	†		†	†
Total	3.40-3.67		6.36- 8.88	12.74-20.82
Grand Total	7.59-7.86		23.18-26.36	33.68-44.49

Source: *The economic value of the ocean resources to the United States*, National Ocean Policy Study, Senate Committee on Commerce, 93rd Congress, 2nd Sess., 1974.

*Insignificant.

†Potentially significant, but unmeasurable.

and gravel, through heavy minerals associated with beach deposits, to surface deposits of manganese and phosphorite.[4] Resources that are presently exploited, and those that will undoubtedly be exploited in the future, include sand, gravel, gold, tin, platinum, diamonds, titanium minerals ileminite and rutile, phosphorite, manganese nodules, zinc, copper, lead, and silver.[5]

Manganese nodules, with their mineral content of cobalt, nickel, copper, and manganese, are currently the most sought-after resource in ocean mining. Successful deep seabed mining operations have the economic potential of $200 million per year,[6] in addition to securing access to scarce land-based resources like manganese and copper. Twelve countries have developed or are developing the technology of mining these minerals. Their present primary operating obstacle is the legal jurisdiction to mine resources that lie beyond any coastal state's jurisdiction.

Nonextractive Uses

The world gross tonnage of cargo carried has continued to increase through the last decade, and it is predicted that volume will continue to grow at a rate of 5 to 8 percent per annum[7] from the level of 225 million gross revenue tons in 1971.[8] The changes in maritime industry will occur largely in the shift of products carried and the means by which they are carried. Liquefied natural gas will have its own specially designed tankers to haul the cargo at –260°F, while petroleum is being shipped in consistently larger vessels that now range up to 474,000 dwt. However, the growth in vessel size is currently leveling off. Other technological innovations have brought the Lighter Aboard Ship (LASH), Roll on-Roll off, containerization, and iron ore slurry ships. Submarine petroleum tankers have been proposed as well. Generally the volume of voyages and cargo carried will continue to increase well into the future. The change will evolve around the world fleet's increase in specialization, sophistication, size, and speed.

As a consequence, the ports and harbors of the world are changing their design, depth, and capacity to service present and future generations of ships. Deepwater ports with depths greater than 95 ft will be necessary, as well as specialized facilities for different commodities, maintenance and dry-docking facilities, warehousing, and traffic control systems for the harbor vicinity. The ability to adapt to the maritime industry's needs will largely determine the traffic of cargo for any successful harbor in the future.

Power generation from the oceans (other than OTEC) has long been used and envisioned by public utilities. Currently, electric generating facilities use coastal waters for their cooling systems, while industrialists and academics have proposed offshore nuclear and fossil-fuel plants. These present uses are limiting in that they view the oceans as either a heat sink or a locale for removing nuisance industries from valuable coastal real estate.

The oceans exist as the major receptacle for the waste from onshore activities. Among these substances are industrial, municipal, and agricultural wastes that are disposed of in the world's coastal waters. Presently restrictions on radioactive waste disposal and other hazardous material exist, and their disposal in the oceans is being curtailed. The issue for the future is the volume, composition, and treatment of wastes before they are dumped in the ocean.

Marine recreation has realized a tremendous growth over the past decade in terms of the number of recreational activities and individual participation. The attraction of the ocean is that it offers a breadth of opportunities—from active water sports to passive sight-seeing. The oceans will undoubtedly continue to entertain more activities and individuals in the future.

New construction of offshore ports, islands, reefs, airports, and industrial complexes has been envisioned.

Nonuse of the marine environment should also exist as a valid consideration. Estuarine sanctuaries, underwater parks, and scenic vistas all comprise nonuses of the marine environment that should be carefully weighed in the allocation and management of the oceans.

The utilization of the ocean and its resources is clearly a function of basic human onshore needs. It can be asserted that as the world population increases, so will the effort to exploit the oceans' resources, and many nations possess either the technological capability or purchasing power to actively seek out these resources. Future activities in the ocean will be of two types: (1) existing activities with higher degrees of efficiency and sophistication, or (2) new uses that previously did not have the technology, necessity, or enlightenment to be operational.

Possible Conflicts of OTEC with Ocean Space Uses

The previous section discussed the extent to which the use of the ocean environment, primarily for human land-based needs, will increase because of both existing and emerging activities. The predicted volume of activity will undoubtedly create conflicts as to the proper, wise, and safe use of the ocean environment. OTEC represents one such emerging use, and its normal operations may conflict with no less than seven other uses. Specifically, OTEC may conflict with aspects of navigation, fishing, waste disposal, recreational boating, LNG conversion, military operations, and offshore mining or exploration. It should be noted that this list does not concern itself with potential environmental conflicts mentioned in Chapters 7 and 10; nor does this list necessarily cover all the potential conflicts—only those of major impact. This section will present a discussion of possible conflicts, with the issue of resolution of conflicts to follow in the next section.

In general, siting will be the ultimate source of conflict surrounding OTEC, and it should be realized that the siting of an OTEC is dependent upon two dominant factors: (1) the thermal gradient and (2) the type of energy use. First, as mentioned in Chapter 1, an OTEC device needs to be located in those environments that possess a maximum thermal gradient ($\Delta T°$), and such areas are limited to the Gulf Stream and tropical regions. Second, the energy produced needs to be: (1) transmitted to shore via cable; (2) converted into other forms of energy (hydrogen) and then piped or shipped to shore; or (3) consumed at sea by energy-intensive industry. The current technical transmission capability of option 1 is approximately 100 mi, and the onshore terminal must be within reasonable access to the power grid. With the exception of superconductive cabling being developed, options 2 and 3 are independent of this requirement. Superconductive cabling is the possible transmission of energy by using certain nonmagnetic metals at cooled temperature with no resistance, voltage loss, or heating produced.[9]

These technical facts explain the primary determinants in OTEC siting, for it is the issue of OTEC siting that ultimately creates most conflicts with other ocean space uses. However, each OTEC site will have a unique set of circumstances that will have to be weighed against the merits of the activities.

The most likely areas of conflict with OTEC would include the following.

Navigation

Using existing technology, an OTEC device sited under option 1 would be located within 100 mi of the coast, an area where the concentration of vessel traffic is greatest. If we set aside the issue of navigational aids temporarily, the frequency of navigational collisions and incidents increases logarithmically with proximity to shore.[10]

Distance from Shore	*Percent of Accidents*
Less than 5 mi	85.2
5 to 25 mi	10.0
Greater than 25 mi	3.3
Not known	1.5

These accidents occur primarily at "choke points" of navigation, i.e., harbor approaches or international straits. If an OTEC is located within one of these choke points, the probability of a collision is greatly increased.

It is noted in Chapter 6 that international and domestic regulations for a novel offshore structure such as this are either unaddressed by international bodies or below optimal standards. This lack of organization poses a real possibility of inadequate navigational aids being installed on OTEC devices. Furthermore, U.S. Coast Guard regulations for offshore drilling rigs (which are 80 to 100 ft above sea level) would not necessarily be optimal for submerged or semisubmerged OTEC devices.

The inadequacy in charting navigational hazards has been a historical condition that remains today despite the efforts of the International Hydrographic Office (IHO). The major problems are undersubscription to IHO and the failure to purchase revised nautical charts. If there were to be a proliferation of OTEC devices between 1985 and 2000, the IHO or any other organization most likely would be unable to chart and disseminate information on these locations in a timely manner if current conditions persist.

Three subsurface extensions of an OTEC device—the cold-water pipe, the mooring, and the transmission cables under option 1—are perceived as additional conflict-producing aspects of OTEC. The cold-water pipe and mooring cable exist as potential hazards to civilian and military submarine traffic, and the transmission cable could possibly transect anchoring areas of coastal tankers.

Navigational hazards need to be seriously regarded in connection with protection of life and property at sea and the maintenance of freedom of navigation. Safety and recognition of international law require that the preceding conflicts involving navigation be given attention and resolution.

Fishing

Conflicts with traditional fishing activities at potential OTEC sites depend primarily upon the facilities being located within 200 mi of the coast. Due to the immense phytoplankton content of these coastal environments, plus estuaries and streams as habitats for numerous anadromous and pelagic species, 93 percent of the world catch occurs within 200 mi of a coast.[11] The other 7 percent occurs primarily at substantial distances from the coast in the east and west Pacific.[12] It should be noted that the tropical coastal regions where OTEC would operate presently have limited fishing activity.

The conflict between OTEC and coastal fishermen occurs because of fishing gear used, adequacy of navigational aids, and potential disruption of traditional fishing areas. Difficulty may also arise between bottom trawlers and the OTEC transmission cable and its anchoring platform.[13]

If artificial upwelling does in fact occur at the bottom water intake, it is conceivable that coastal species and some pelagic species would be attracted to an OTEC facility. Should this be the case, an issue arises concerning the type of fishing activity which could be performed within an acceptable distance from an OTEC device without additional risk.

Conversely, objections by fishermen have been lodged in the past regarding oil rigs as a disruption of historical fishing activity. Industry has claimed that such structures increase the amount of fish in the area by acting as a "habitat."[14] The inexactness of the science and the variability of fish populations leave this argument unresolved. Therefore, it is reasonable to ask two questions: If offshore structures are disruptive, do fishermen have a right to compensation? And if they lose an expensive dragging net in OTEC equipment, should they be compensated for their losses?

Offshore Mining and Exploration

The U.S. Department of the Interior is empowered to lease areas of the outer continental shelf (OCS) for exploration and exploitation of the shelf and subsoil.[15] The current law-of-the-sea conference is negotiating an international regime to regulate the exploration and exploitation of the seabed and subsoil beyond national jurisdiction. Both current national and international law deny claims of sovereign rights to the waters superjacent to claimed or leased areas.[16]

Consequently, it would appear that OTEC would have the legal right to operate in the waters superjacent to any nationally or internationally leased seabed or subsoil.

Spatial problems might arise if offshore oil or mining operations are conducted within proximity to an OTEC device. Conflicts might arise from these operations' effluents or operation of their service vessels within the area. In addition, it is calculated that if an OTEC is moored at 20,000 ft in a 3-kn surface current, the plant would rotate about a 7800-ft-radius circle.[17]

Mutually Exclusive Areas or Activities

There are two possible types of ocean uses that would prevent an OTEC plant from operating in proximity to them. The first is an LNG conversion plant which transforms LNG from a liquid to a gaseous state with a resultant thermal discharge into the oceans at temperatures below ambient temperature. The plume and impact area of this discharge is not readily known (thermal discharges in coastal harbors are currently predicted to be 12°F below ambient temperature within 0.5 mi of the discharge), but it is theorized that an LNG conversion plant could lower the surface water temperature intake. Such lowered ΔT° would lessen the already extremely low efficiency of the OTEC plant.

A second technical problem concerns the relationship between municipal waste disposal and biological productivity. True municipal (not industrial) waste, when only primarily treated, contains a high degree of nutrients which consequently increases the biological productivity of receiving waters. In such cases, any increase in biological productivity could increase the degree of biofouling of the heat exchangers in an OTEC plant. Biofouling of the heat exchangers would directly decrease the efficiency of a process that cannot afford such losses.

These two situations—LNG and municipal waste disposal—are examples of why OTEC plant efficiency must be considered in the siting of ocean space uses, regardless of the remoteness of the possibilities. In dealing with plant efficiency factors of only 2 to 5 percent, additional losses cannot be afforded.

Military Uses

The most difficult of all conflicts to predict is that concerning military operations. However, OTEC plants might conflict with military operations if they obstructed military maneuvers, weapons testing, long-range missile reentry, or devices and installations for tactical and strategic purposes.[18]

Two considerations are involved in siting structures offshore: (1) The Department of Defense (DOD) historically has taken preemptive actions in the approval of offshore mining and exploration, and (2) many DOD activities are classified, which makes plans or responses difficult. Two examples concerning DOD and offshore operations are worthy of discussion. As noted in Chapter 9, the President has established numerous defensive sea areas adjacent to the coast in which siting of fixed structures and some traffic are prohibited.

The history of phosphorite mining offshore California demonstrates such a situation.[19] The Collier Corporation was interested in phosphorite deposits on the outer continental shelf off southern California, and the Department of Interior was inclined to lease the offshore area. The Department of Defense desired to have the right to have the mining operations halted and equipment removed at the lessee's expense if, in the judgment of DOD, national security required such action. These stipulations were in conflict with Section 12(c) of the Outer Continental Shelf Lands Act.[20]

> All leases issued under this Act, and leases, the maintenance and operation of which are authorized under this Act, shall contain or be construed to contain a provision whereby authority is vested in the Secretary (of the Interior) upon a recommendation of the Secretary of Defense, during a state of war or national emergency declared by the Congress or the President of the United States after the effective date of this Act, to suspend operations under any lease; and all such leases shall contain or be construed to contain provisions for the payment of just compensation to the lessee whose operations are thus suspended.

DOD ultimately prevailed, and consequently the resulting high degree of risk of investment removed the desire of Collier Corporation to mine offshore. It is possible that prime ΔT° and energy consumer areas may also be designated as defensive sea areas.

Another concern of DOD is the increasing volume of acoustical noise in the ocean environment, since the main military interest in the seabed and the deep ocean lies in the context of antisubmarine warfare.[21] The majority of antisubmarine detection and surveillance is accomplished by acoustical sensing and tracking. Such operations are hindered by artificial acoustical noise in the ocean environment. OTEC operations represent an addition to this volume, and it is conceivable that by 1985 or 2000 DOD might show greater concern for the interference in the sensing and tracking of foreign submarines. The preemptive power of DOD should not be overlooked or underestimated when OTEC siting issues are discussed.

Other Considerations

Potential conflicts might occur if OTEC is used as the power supply for floating military platforms, ocean mining operations, or scientific research. The current law-of-the-sea conference is seeking to restrict and control ocean mining and scientific research, as well as emphasizing peaceful uses of the sea. Thus, through association, OTEC might become subject to additional control or restriction if the plant were beyond the limits of national jurisdiction.

Summary

The two dominant factors concerning the siting of an OTEC plant—the ΔT° and the use of energy—will inevitably create conflicts with other ocean space uses. The preceding discussion has enumerated the potential conflicts, each with a varying degree of severity. These identifiable conflicts may be resolved by applying or amending existing legislation, drafting new legislation, or by producing technology that would eliminate the conflict.

Resolution of Conflict

The resolution of conflicts between OTEC devices and other ocean space uses is particularly difficult for two reasons. (1) Theoretically an OTEC plant could be located in the waters and jurisdiction of the United States, the superjacent waters of the outer continental shelf, the high seas, or the jurisdiction of another state. (2) The current juridical status and boundaries of these zones, plus the proposed economic zone, are presently being negotiated by the law-of-the-sea conference which may substantially revise the current state of affairs.

This section will discuss how resolution of conflicts might be achieved through technology; government intervention, negotiation, and arbitration; or legal action. In principle, it is my opinion that the United States, the major proponent of this technology, should establish standards that will earn the respect and approval of the international community.

It may appear that this chapter endorses the current draft of the Informal Single Negotiating Text (ISNT) produced at the 1975 Geneva session of the law-of-the-sea conference. This is true only as it applies to OTEC, for the current draft provides substantial gains in articulation of avoiding and resolving emerging use conflicts. Without speculating on the adoption of ISNT, it does provide valid indications of the state of affairs from 1985 to 2000.

Resolution of Conflicts within U.S. Jurisdiction

As noted in Chapter 3, within internal waters and territorial seas the jurisdiction of the coastal state is near absolute—with exceptions existing only for innocent passage and entry in distress. The proposed 200-mi exclusive economic zone would establish for the coastal state, in the words of ISNT, "sovereign rights for the purpose of exploring and exploiting, conserving and managing the natural resources—whether renewable or non-renewable—of the bed and subsoil and the *superjacent waters*" [emphasis added].[22] Specifically, it would establish "exclusive jurisdiction with regard to other activities for the economic exploitation and exploration of the zone, *such as production of energy from the water*, currents and winds . . . " [emphasis added].[23] It is, therefore, evident that the United States possesses near absolute power over its territorial sea and probably limited sovereign rights over an exclusive economic zone. This likely definition of U.S. jurisdiction is appropriate for the installation of an OTEC plant for the purpose of transmission of electric power via submarine cable to shore.

Within the U.S. jurisdiction, a heavy emphasis should be placed on the advancement of technology to resolve potential conflicts. Where such technology fails to either materialize or be implemented, legislation, treaties, conciliation, or arbitration should be sought as necessary remedies for conflict.

In addition, the need for a U.S. lead agency for regulation and licensing will be a recurring recommendation of this section, for due to the number of affected agencies there can be no substitute for efficient interagency coordination.

Navigation

Assuming that any OTEC site is chosen on the basis of ΔT° and the proximity to the power grid, navigational conflicts will inevitably occur. The following methods of resolution should be considered.

Shipping lanes should be avoided, perhaps through the use of buffer zones as proposed by Offshore Systems.[24] Safety zones could also be established pursuant to provisions of the Convention on the Continental Shelf[25] and Article 48, Part III, of ISNT.[26] Under this approach, a 500-m safety perimeter would be established and required to be respected by the ships of all nations.

For those OTEC plants located on the high seas as a "reasonable use"[27] or as a "freedom to construct artificial islands and other installations permitted under international law,"[28] 500-m safety zone perimeters should be created for the purpose of safety. Plants located this distance from shore probably will not be involved in direct transmission of electricity to the coast. Consequently, international regulation of industrial effluents into the marine environment will be necessary. However, it is possible that other nations will perceive this safety perimeter as either an "industrialization" or "colonization" of the high seas. The next section will deal with how such interpretations can be clarified.

Coastal states should be empowered to impose navigational standards to be acknowledged by ships of all nations which pass through OTEC areas. Currently, the U.S. Coast Guard has the responsibility for aids to navigation,[29] and that agency should establish navigational standards similar to those applying to oil rigs, which would reduce risk of loss of property and life to the lowest possible level.[30] Initially, the U.S. Coast Guard should institute the equivalent of a "code of conduct" for use in OTEC navigational devices. This should include provisions to increase radar and visual detection, as well as radio communication. Consideration should also be given to an OTEC plant serving as a mariner's aid to navigation.[31] This proposed code should be mandatory for all OTEC devices within U.S. jurisdiction. Where a lack of international standards exists, this code could serve as the standard for vessels under its flag as per international law for "the use of signals, the maintenance of communications and the prevention of collisions."[32] By ensuring prompt and adequate navigational aid installation, the United States could encourage IMCO to adopt similar standards for all other OTEC devices located on the high seas. As noted in Chapter 6:

> Based upon this practice, it is entirely possible that within the framework of IMCO machinery, there could be developed for OTEC devices a code of recommended practice that in time, based upon actual experience, could be reduced to treaty form.

Since OTEC is being proposed primarily by one country at this time, it is doubtful that IMCO would place a high enough priority on OTEC to generate a treaty. Therefore, it appears clearly within the purview of the U.S. Coast Guard to take the initiative.

Charting can be adequately performed if the International Hydrographic Organization (IHO) is given advance notice of OTEC sites, safety zone size, and so forth. Then this information could be incorporated into the navigational charts which are updated regularly. It is currently being proposed that regional or "block" updates of charts be revised and circulated more frequently, rather than have the traditional dissemination of international series.[33] This effort would decrease the existence of outdated charts that do not take into account new navigational hazards.

A part of any lead agency's responsibility should be the requirement of notifying IHO of the location and timing of proposed OTEC sites. If one assumes a lag time between procurement of all licenses and actual construction and installation, this period should be adequate to ensure the proper notation of OTEC plants as navigational hazards.

Although maritime safety regarding civilian and military submarine activities has been ignored in the past, OTEC devices probably should have sonar inducers located in their underwater structures, particularly in the mooring cable.

Cable laying and the protection of cables are generally covered by existing law and international agreement,[34] as well as being considered in proposed international agreements.[35] Those cables lying on the continental shelf will probably be subject to OCS mining and exploration regulations. Caution and regulation would thus be necessary to avoid conflict and damage to OTEC transmission cables or pipelines.

It was assumed that all OTEC plants would be stationary and not permitted to "graze" for $\Delta T°$. If an OTEC device were permitted to graze, navigational conflicts would be extremely difficult to resolve, even though there are provisions for ships at rest and ships under command but not underway.[36]

Fishing

Fishing and OTEC conflicts are more difficult to rationalize or resolve due to the ambiguity of both the rights of fishermen and the true impact of OTEC. Any negative impact on fishing activities or traditional fishing areas should be noted in the draft environmental impact statement required under the National Environmental Protection Act.[37] These impacts would then be factors in permit issuance from the lead agency. Opportunities for resolution are limited outside of NEPA.

The issue of gear loss caused by OTEC cables, moorings, or anchors creates another area of consideration for new legislation. Liability laws instituted to protect fishermen from these losses would substantially ameliorate their attitude toward such a device.[38]

Regarding OTEC devices beyond U.S. jurisdiction, reasonable regard for the freedom of fishing of all states would have to be exercised.[39] Referring to the adequacy of international law on the subject, E.A. Brown has noted that "[t]hese very general rules offer no easy solutions to conflicts of interest except in extreme cases where, for example, the expected value of the Shelf resource clearly outweighs that of minority fishing interest."[40]

Mutually Exclusive Uses

In the cases of LNG and municipal waste, both are largely domestic issues.[41] Should a future situation occur in which an OTEC site is in proximity to a municipal waste disposal site, one should consider the eventual impact of Environmental Protection Agency standards for effluents. Secondary waste disposal is required by 1977, and the best available technology (possibly tertiary) by 1983.[42] It is arguable that these standards will substantially reduce the nutrient contents of the effluents and consequently reduce biological productivity. However, it is doubtful that these risks would be assumed or that alternative locations would be sought.

The decision to avoid these two environments is largely one of the corporate or decisionmaking body. However, if a site is selected and licensed in absence of these two activities, or any other conflicting activity, other licensing agencies should be expected to maintain this exclusive environment. If two mutually exclusive sites were simultaneously requested for the same vicinity, the licensing authority might invoke the site-dependency concept. *Site dependency* is the evaluation of the imperativeness of any activity in that location on the basis of physical environment, economics, and social necessity.

Military Uses

Wherever possible, site selection should avoid exclusive defense zones. However, where unavoidable, the lead agency for licensing should engage in affirmative interaction with DOD in securing the preferred site and guaranteeing the ability of the OTEC plant to operate there for the duration of the license, lease, or permit. Interagency agreements to accommodate such conflicts have been developed in the past[43] and would be necessary with OTEC as well. Investors of private capital probably will avoid any uncertainty dealing with tenure of site availability. Such defense areas should stand the test of site dependency against all other ocean uses.

Offshore Mining and Exploration

It is possible that most of the U.S. continental shelf will be up for bid on oil and gas exploration and exploitation over the next 10 years. Such accelerated lease schedules will probably place OCS development and OTEC operations in the same area, i.e., the Gulf of Mexico and the southeast Atlantic. Although the leases are for the seafloor and subsoil and not the superjacent waters, the rigs, supply vessels, etc., will impact any OTEC within the vicinity.

Therefore, it is necessary to create legislation that provides authority over spatial distribution of structures in OCS superjacent waters within U.S. jurisdiction. Additionally, similar management is needed in laying submarine cables and pipelines to avoid injury and damage to such objects or obstruction to mining and exploration activities. This is necessary over and above the obstruction-to-navigable-waters function of the U.S. Army Corps of Engineers and other OCS lease terms. The probable increase in all types of OCS activity necessitates such legislation.

For mining and exploration of the seabed and subsoil beyond national jurisdiction, the matter is ill defined. The current negotiating text provides for only the area leased—not the superjacent waters—and reasonable regard must be paid to the interest of other states in their exercise of freedom of the high seas. Beyond the 500-m safety zone, it is agreed that no state has the legal right to restrict the navigation on the high seas.

Summary

The resolution of conflicts is difficult to prescribe because existing legislation, international agreements, and customary international law do not address the

OTEC concept. In most cases, "reasonable" and "equitable" conduct is advocated; however, various interpretations of actions immediately create real or imagined conflicts. Obviously it is important to develop the technology that will avoid many of these spatial conflicts, e.g., navigational aids and protection of submarine cables.

However, it is far easier to propose technological developments and new legislation than to instigate or implement them. Accordingly, the following section will discuss the means available to resolve conflict—alternatives that the United States and the international community have in articulating appropriate conduct for emerging uses of the sea.

Means of Resolving Conflicts

This section does not undertake the presumptuous task of putting forth all options, avenues, or philosophies of conflict resolution. Rather, I have sought to provide viable alternatives for dealing with the emergence of OTEC. The manner in which OTEC development is managed may be a test of the adequacy of domestic and international laws in dealing with other emerging uses of the sea. As Edward Wenk put it:

> What steps are necessary for the rational management of ocean space to maximize the effective use of the limited resources, to distribute the benefits equitably and rationally, and to consider the long term dilemmas as well as short term needs?[44]

Domestic Options

Since the first OTEC plant will not be online until 1987-1989, the lead time afforded would be wisely spent in achieving a clear articulation of public policy regarding OTEC. In situations where lead time does not exist, the courts often act as the primary, if not the sole, responsive branch of government. It is difficult to believe that the judicial system is the most competent or responsible party for the establishment of a marine resources management system. The mechanisms for establishing laws offer the ability to define and articulate policy, whereas the courts might best serve in interpreting the policy and settling claims or conflicts.

In many cases existing legislation provides various parties with predictability of law and planning with regard to OTEC; i.e., regulatory agencies have clear and adequate authority over various aspects of subsystems of OTEC. In these cases, it would be necessary for such agencies to promulgate regulations in accordance with enabling legislation. Where agencies fail to exercise their responsibilities, suits to enjoin them to do so would be necessary.

If a regulatory agency does not have clear authority and existing legislation is deemed inadequate to meet the needs of OTEC, amendment of that legislation may be appropriate. These actions recognize a preexisting authority, as well as an historical evaluation of performance or exertion of authority by the regulatory agency.

If neither of these alternatives satisfactorily meets the institutional and legal needs of OTEC, creation of new legislation remains. Several basic questions immediately surface: Has the Deepwater Ports Act of 1974 set a precedent in requiring new and comprehensive legislation for each emerging use of ocean space? Is broad legislation capable of sufficient articulation of policy and delegation of authority for all emerging ocean space uses?

At a minimum, lead agency designation is imperative for the commencement of OTEC planning and licensing. A broader issue would deal with attempts to establish a department of energy and natural resources, a single ocean agency, or a national energy siting act. This simplified approach is the rational means of resolving conflicts. Furthermore, it forces the issue as to what level and body of government should be responsible for the establishment of ocean policy.

International Options

The current law-of-the-sea conference is attempting to establish policy, jurisdiction, and institutions regarding the development and wise use of ocean resources. Even if the conference is successful in adopting a treaty, it is realized that the treaty in itself will not be the last effort in ocean resources management. Delegates will need to continually reconvene to resolve and amend the treaty, primarily as it pertains to emerging uses of the sea. If OTEC becomes a reality and its energy is consumed at sea, then it will be necessary for the international community to address the permissibility of power plants and industrial complexes beyond the extent of national jurisdiction.

Notes

1. Pardo, *Perspectives on the Law of the Sea Negotiations* (1975).

2. Dempster, "The Role of Science and Technology in the Revitalization of Our Fishing Industry," in *The Oceans and National Economic Development*, U.S. Senate Commerce Committee (1973) at 112.

3. Doumani, *Ocean Wealth, Policy and Potential* (1973) at 28.

4. Id., at 20.

5. Id., at 20-31.

6. Moncreiff and Adams, "The Economics of First Generation Manganese Nodule Operations," *Mining Congress Journal* 49 (December 1974).

7. Brown and Cooper, "Future Shipping and Transport Technology and Its Impact on Law of the Seas," in Clingan, ed., *Law of the Sea: Caracas and Beyond* (1973) at 272.

8. Frankel, *Demand and Supply of Shipping, A World Review*, Report no. 70-12, MIT Commodity Transportation and Economic Development Laboratory (June 1972) at 25.

9. Scurlock, "Superconductors, A New Technique for the Power Industry," *Electronics & Power* 524 (11 July 1974).

10. Crimes, "Survey of Marine Accidents," 25 *Navigational Journal* 4, 502.

11. *World Fishery Resources*, United Nations, FAO (1972).

12. Id.

13. Schafer, *Proceedings of Conference on Law Organization and Security in the Use of the Ocean*, University of Ohio (1967) at 14.

14. Treybig, "How Offshore Platforms Help Fishing," *Ocean Industry* (April 1971), *passim.*

15. Outer Continental Shelf Lands Act, 43 U.S.C. §§ 1331-43 (1964) (originally enacted as Act of August 7, 1953, ch. 345, 67 Stat. 462).

16. Brown, *The Legal Regime of Hydrospace* (1971) at 91.

17. Trimble, "OTEC System Study Report," *Proceedings 3rd Workshop on OTEC* (May, 1975) at 13.

18. Goldblat, "Law of the Sea and the Security of Coastal States," in Clingan, ed., *Law of the Sea: Caracas and Beyond* (1973), pp. 301-24.

19. Juda, *Ocean Space Rights* (1975) at 66-67.

20. *Supra* note 15.

21. Hirdman, "Military Interests to be Negotiated," in Alexander, ed., *The Law of the Sea: A New Geneva Conference* (1971) at 89.

22. ISNT, U.N. Doc. A/Conf. 62/WP.8/Part II, Art. 45(1)(a).

23. Id., Art. 45(1)(1)(i).

24. Kehnemuyi and Lochbaum, "Offshore Nuclear Power Plants," *Marine Technology* 251 (July 1973).

25. Convention on the Continental Shelf [*done* April 29, 1958, 15 U.S.T. 471 (1964), T.I.A.S. No. 5578, 499 U.N.T.S. 311, *in force* June 10, 1965], Art. 5.

26. ISNT, U.N. Doc. A/Conf.62/WP.8/Part III, Art. 48.

27. *See* Chapter 3.

28. ISNT, *supra* note 22, Art. 75(1)(d).

29. *See* 14 U.S.C. 81.

30. *See* 33 C.F.R. 62.

31. Id.

32. ISNT, *supra* note 22, Art. 80(3)(c).

33. Pascoe, "International Charts," 25 *Journal of Navigation* (No. 2) 258.

34. Convention on the High Seas, Arts. 2, 26-29.

35. ISNT, *supra* note 22, Arts. 18(c), 47, 48, 65, 75(c), and 99-102(c).

36. *See, e.g.*, 33 U.S.C. 1064, and the Convention for the Safety of Life at Sea.

37. National Environmental Policy Act of 1970, 42 U.S.C. § 4321, *et seq.*

38. Such agreements were made, for example, between Cook Inlet (Alaska) crab fishermen and oil companies for the approval of exploratory drilling.

39. Convention on the High Seas, Art. 2.

40. Brown, *The Legal Regime of Hydrospace* (1971) at 96.

41. There are no discharges of municipal wastes 200 mi out to sea, and it is unlikely that an LNG conversion plant would be located that far offshore due to the cost of piping it onshore.

42. *See* 33 U.S.C. §§ 1251-1376.

43. *See* "Oil and Gas Lease Sale," 38 *Fed. Reg.* 222 at 31,855, ¶ 10(2).

44. Wenk, *The Politics of the Ocean* (1972) at 428.

International Regulatory Authority concerning Ocean Thermal Energy Conversion Devices

C.R. Hallberg

Introduction

There is no international organization either within or without the United Nations system, which can be said to be clearly in possession of the authority or competence to deal with OTEC devices per se. There have been proposals set forth during the negotiations on the law of the sea contemplating the creation of an international regulatory mechanism for seabed and ocean resources management. The terms of reference of such a regulatory mechanism have yet to be developed in any detail, and it would be difficult to forecast whether the regulation of OTEC devices would or could fall within its purview. Certainly the preparatory work done to date in the law-of-the-sea negotiations gives little evidence that any serious consideration has been given to the extraction of energy from the ocean or superjacent atmosphere as matters within the competence of such an organization.

An examination of the extant international organizations indicates that OTEC devices could be considered within the purview of at least two specialized agencies of the United Nations, but the same examination will reveal that many kinds of maritime devices are essentially unaddressed by any international organization. The necessity for establishing internationally agreed-upon standards relating to OTEC devices is probably subordinate in priority to the necessity of developing norms for certain existing maritime devices. The extent to which any or all of these devices may be subject to consideration within an intergovernmental or international body is largely dependent upon whether the philosophical approach by the world community to maritime matters continues in its present form or is modified.

Background

The three traditional and principal uses of the sea have been subject to some degree of international accord for a period of years. Two of these uses have lead to the development of international bodies of a more permanent nature than the customary ad hoc conference which has been employed to elaborate conventions. Being an exercise of sovereignty, the third major use—naval—has not inspired the same sort of organizational approach.

The regulation of fishing is a contentious matter, and as a result there has been no general accord among nations on this topic. However, with regard to certain fish stocks or to certain fishing grounds, several international bodies of a continuing nature, but with fairly limited memberships and even more limited authority, have been created. The International Whaling Commission and the Inter-American Tropical Tuna Commission are but two examples of this species of international body. Since regulatory bodies created for fisheries purposes have as their reason for being the allocation and conservation of fish stocks rather than a broader purpose, which would include the relationship of the fisheries to other ocean uses, it does not appear that fisheries-oriented international entities would have any application to OTEC regulation.

Commercial shipping has been subjected to the broadest range of international regulation of any ocean activity. For the most part, this regulation has taken the form of internationally agreed-upon standards implemented by domestic law of the party nations and enforced by the parties without the involvement of any international body. Examples of this procedure, which is most pertinent in the commercial and liability aspects of shipping, include a series of so-called Brussels treaties: the 1910 Collision, the 1910 Assistance and Salvage, the 1926 Liens and Mortgages, and the 1957 Limitation of Shipowner's Liability.

Concerning the regulation of maritime safety matters, several different techniques have developed, but the genesis of nearly all the concepts involved was the 1889 Washington Maritime Conference. In terms of concrete achievements, the conference established the first truly international set of collision rules as well as ship-routing regulations for the North Atlantic. These regulations, as with the commercial treaties, were enforced without the involvement of any international body. The concept of an international body for the exchange of information and the unification of maritime law was first broached at this conference. The initial step consisted of dealing with the exchange of hydrographic information, emphasized in the St. Petersburg Navigation Congresses of 1908 and 1912, culminating in the 1919 London International Hydrographic Conference which created the International Hydrographic Bureau. This body is quite active in the exchange of all kinds of marine hydrographic information, but it suffers from the obvious disadvantage that only a few of the larger nations can afford the costs of a hydrographic survey organization, a chart-making facility, and the extensive radio communications facility required to disseminate notices to mariners. Obviously, to the extent that OTEC devices may present an interference to the other uses of the sea, there will be an involvement with IHO. However, the limited scope of this body would not make it suitable for any general regulatory or advisory responsibility.

The one remaining vestige of the 1919 Treaty of Versailles is the International Labor Organization. It is entirely a consultative and recommendatory body. This aspect facilitates member states in agreeing to the various resolutions this body has formulated over the years since they are not binding upon states.

Maritime labor matters have been rather sparingly treated, and the Officers Competency Convention of 1936 represents the major piece of international legislation in this field. The ILO has not yet considered to any degree the labor situation in the offshore mineral mining area, and it would appear unlikely that any attention will be paid to the labor situation involved in OTEC devices.

The principal international maritime safety legislation—the Safety of Life at Sea Convention—had its genesis in the ill-starred 1914 London conference. The first in-force convention was elaborated in 1929, followed by a second generation in 1948. Neither of these involved an international body. The same was true of the first Load Line Convention (which is deceptively entitled since its principal purpose is to establish structural parameters for ship construction). Since the later editions of these conventions have been elaborated under the aegis of the Intergovernmental Maritime Consultative Organization, further discussion of these instruments will be subsumed in the discussion of IMCO.

With regard to aids to navigation, such as buoys, lights, lightships, and electronic devices, there have been no international agreements (except two of minor significance); consequently, there is no international regulatory body. However, the various buoyage authorities in the world are associated in the International Association of Lighthouse Authorities, founded in 1929 and reorganized in 1957. The IALA serves as a medium of exchange of information, principally of a pharological nature. It is not an association of states. To the extent that OTEC devices might be required to display warning signals, IALA would be a useful consultative body with regard to the most efficacious kinds of signals; but beyond that its organizational nature would not lend itself to regulation of OTEC devices.

In recognition of the need to concentrate the focus of world attention on oceanographic research in one international body, the Intergovernmental Oceanographic Commission (IOC) was created in UNESCO. While principally involved in the coordination of joint oceanographic ventures, the IOC has turned its attention to the peculiar problems involved in the deployment of devices, principally unstaffed, used for the collection of ocean data. These devices, carrying the acronym ODAS (Ocean Data Acquisition Systems and Platforms), have many common characteristics with prospective OTEC devices.

In collaboration with IMCO, recommended markings and signals for ODAS were developed and promulgated in the early 1960s. However, in recognition of the larger scope of problems relating to the deployment and status of ODAS, the IOC undertook the elaboration of an international instrument to treat the matter. To date the attempts to deal with the so-called public law side of the situation, focusing on rights of deployment in various ocean areas, have been frustrated because of the ongoing law-of-the-sea negotiations which have subsumed many of the issues. The private law side has just begun to be addressed, and it is unlikely that there will be any early resolution of these issues. Since the underlying motivation of the IOC is exploration rather than exploitation, despite the characteristics common to ODAS and OTEC devices, UNESCO-IOC does not appear to be the international body most suited to take cognizance of OTEC devices. It should be observed that in its current draft form the proposed ODAS convention does not contemplate the creation of any continuing international body to deal with ODAS matters.

Finally picking up the initiative of the 1889 Conference, the UN in 1947 established a Provisional Maritime Consultative Council which was followed by the United Nations Maritime Conference in Geneva in 1948, which elaborated the convention on the Intergovernmental Maritime Consultative Organization. This treaty came into force some ten years later.

The basic objectives of the organization were as follows: to provide machinery for cooperation among governments in the field of governmental regulations and practices relating to technical matters of all kinds that affect shipping engaged in international trade; to encourage general adoption of the highest practicable standards in matters concerning maritime safety and efficiency of navigation; to encourage removal of discriminatory action and unnecessary restrictions by governments affecting shipping engaged in international trade so as to promote the availability of shipping services to the commerce of the world without discrimination; to provide for the consideration by the organization of matters concerning unfair restrictive practices of shipping concerns; to provide for consideration by the organization of matters concerning shipping that may be referred to it by any organ or specialized agency of the United Nations; and to provide for the exchange of information among governments on matters under consideration by the organization.

It is fair to say that some maritime states were concerned with the IMCO charter provisions which might give an international body some control over the economics of world shipping, and the delay incurred in the coming into force of the convention was the result of behind-the-scenes negotiations to ensure that IMCO would not get into the Liner Conference business. It is of interest to note that the closest approach within IMCO to this facet of maritime affairs has been in the Facilitation Convention of 1966, which deals principally with easing the red tape of ship and cargo entry.

Pollution of the sea by oil is germane to this organizational analysis only in that the 1953 Oil Pollution Conference, in elaborating the first treaty on this topic, chose the embryonic IMCO as the ultimate depository despite the absence of any specific authority in the IMCO charter to treat pollution matters. There have followed several amendments to the 1953 Convention as well as the elaboration of five other instruments treating various aspects of pollution, all done under the aegis of IMCO.

The Safety of Life at Sea conventions of 1960 and 1974, the Collision Regulations of 1960 and 1972, the Container Safety Convention of 1972, the Loadline Convention of 1966, and several similar instruments have all been developed within the framework of the IMCO machinery. All the foregoing conventions contain regulatory provisions which could have application to OTEC devices.

Installations and devices used for seabed exploration and exploitation, of course, are dealt with to a certain extent in the 1958 Geneva Convention on the Continental Shelf. This instrument treats the right of deployment of the devices as well as the more fundamental issues of the ownership of the seabed resources. It does not treat the majority of the maritime safety matters involved, nor does it provide for any international regulatory body. Because of the commonality of characteristics that continental shelf installations may have with certain OTEC devices, it is of interest to note that the principal work done concerning maritime safety matters for continental shelf installations has been accomplished by IMCO. To date this has consisted of recommended practices for obstruction marking and certain relatively minimal safety considerations. However, considerable effort is presently being expended in IMCO with a view toward the development of a construction code for mobile drilling units.

In the law-of-the-sea negotiations, the issue of regulation of offshore installations, with the focus on resource allocation and pollution, has been the subject of much debate. The possibility exists that an international regulatory body of a continuing nature may be created to treat these matters. It is entirely speculative as to whether this body will be preoccupied with the more mundane safety matters, or whether machinery will be developed whereby the more technically competent body—IMCO—will provide the requisite standards to some extent.

Discussion

Of the many aspects of maritime activities that would appear to lend themselves to international regulation, shipping has been dealt with the most extensively by far, and a specialized agency of the UN has been created as a continuing consultative body to further improve this regulation. The history of IMCO, short as it is, indicates a definite predilection on the part of that agency to broaden the scope of its interests and move into related fields. This is particularly true with regard to environmental protection. The scope of the most recent convention on this topic, the Marine Pollution Convention of 1973, extends to platforms for certain purposes. Of particular significance to OTEC devices is the agenda item on the IMCO Legal Committee's long-term program to develop and define the legal status of novel types of watercraft. While originally aimed at hydrofoils and air-cushion vehicles, the scope of this study has been increased to include mobile drilling units, civil submersibles, and conceivably OTEC devices. (ODAS have been the subject of a separate joint IMCO-IOC effort, but they could well be subsumed into this agenda item.)

Procedurally IMCO has developed a very useful machinery to deal with new concepts and improved methods of regulation or standard setting. Within one of the standing committees of the organization, a new concept is subjected to technical scrutiny (regardless of how much research may be warranted), economic analysis, and political appraisal. Out of this process comes a recommendation from the committee as to the merits and best application of the concept. Customarily, the governing body of IMCO, the Assembly, will formulate this as a "recommended practice" and encourage member states to take the necessary steps to effectuate the recommendation. After a period of time, the recommendation either proves itself in practice or is discovered to be of insufficient value. In the former case, the recommended practice is recast into the form of an amendment to the pertinent convention, and then it becomes a part of the international regulatory machinery.

In the event that the new concept does not readily fit within the framework of one of the existing IMCO conventions—there are now 19 IMCO conventions and 18 major amendments in existence, not all of which are yet in force—a new convention is elaborated.

Based upon this practice it is entirely possible that within the framework of the IMCO machinery, there could be developed for OTEC devices a code of recommended practice that in time, based upon actual experience, could be reduced to treaty form.

However, it should be recognized that several classes of existing marine devices remain as yet essentially unregulated. There appears to be, for example, the need to standardize the various systems of aids to navigation which exist throughout the world. Currently, the differences between these systems can and do cause confusion. Yet there is no movement of any consequence to regulate this category of devices.

After nearly 30 years of experience with continental shelf installations, most safety aspects of these devices are still left to domestic regulation. While work is going on in IMCO to develop a construction code for mobile drilling units, stationary platforms have not been considered. In particular, the matter of the interference of continental shelf installations with other uses of the sea—shipping and fishing for the most part—remains essentially unregulated.

Offshore loading terminals, which number in excess of a hundred at this writing, are treated only in the IMCO oil pollution conventions (and there to a very limited extent), and perhaps in the Territorial Sea Convention of 1958, under the roadstead article (Art. 9), although because of law-of-the-sea overtones, this is a controversial point.

Submarines, both civil and military, have been thoroughly ignored in relation to general maritime safety matters, even extending to collision regulations.

ODAS now are deployed in fairly respectable numbers throughout the oceans of the world. To date, efforts to regularize their deployment and status have been unavailing.

Conclusion

Of all the existing and prospective international machinery that takes the form of a continuing body, IMCO appears to offer the best prospects for the establishment of adequate standards and regulations to govern many of the aspects of the deployment of OTEC devices. IMCO is probably not equipped to treat any of the economic aspects of ocean energy conversion since it has developed no capability to deal with the economic aspects of shipping, despite the charter provisions. Whether an international entity created as the result of the law-of-the-sea negotiations could be seized with the economic aspects remains speculative.

Until OTEC devices become a "force in being," it is unlikely that they will be addressed in any meaningful way in any international body.

7 International Environmental Aspects

Robert E. Stein

Environmental Impact

A review of the international environmental impacts of ocean thermal energy conversion should probably start with an extension of those impacts felt domestically within a country of operation. If the site of an OTEC facility is sufficiently far offshore to put it close to the edge of the territorial sea, or close to the territorial sea (or economic zone) of another nation, the environmental effects of the plant's presence and its operation would be similar to those effects within a particular state. These effects would include physical disruption, air and water pollution due to the construction and maintenance of the device, discharges of waste from workers on the platform, and boats going to and from the facilities. All might well have an effect in a neighboring nation.

Irrespective of the location of an OTEC facility, a number of environmental issues will be raised. Chemicals used for ridding the OTEC pipe of biological material, which could affix itself and cause fouling, could enter the water column and affect plant and animal species therein. The sucking of cold water from near the ocean bottom could cause upwelling—the disturbance of nutrients which lie on the bottom—with effects unknown at present. A preliminary analysis of this problem is presently underway in a study coordinated by the Department of Commerce to assess the problems caused by upwelling as part of a deep ocean mining operation.[1]

The results of studies dealing with biofouling and upwelling specifically in the context of OTEC are not known at this time, although they are underway. It will have to suffice, therefore, to raise the question of these potential effects.[2]

Moreover, although it is too early to fully appreciate their effect, the longer and broader range of impacts of OTEC technology can be anticipated or at least should be considered. It is possible that placement of a number of OTEC plants along the Gulf Stream on the Eastern Coast of the United States (between Florida and Cape Hatteras) may result in an alteration of the temperature distribution of the water of the Gulf Stream, which could affect fish life nearby and climates as far away as northern Europe. Without any prejudgment of the outcome of a review, the potential climatic effects should be carefully investigated and evaluated. A breach of a working fluid or an accident involving it could also have effects beyond the launching and operating nation. The discharge of a working fluid such as a fluorocarbon compound, which may reduce the earth's protective ozone layer, will certainly be of direct concern to the international community. Moreover, the placement of a series of OTEC plants may have an effect on fish life and patterns of fish migration which might also be of interest to other states traditionally fishing those stocks. As has already been noted, pumps will be required to move large quantities of water from the bottom layer of the ocean, and little is known about the potential effects of moving organisms from this layer to suspension near the surface. The effects just noted presume the configurations of OTEC facilities set forth in Chapter 1, which are designed to produce electricity. If a more distant OTEC plant were used to produce ammonia or hydrogen, or to process at sea minerals mined in the deep ocean, there would be additional effects to be considered. These are discussed in the section on Distant Uses of OTEC.

Environmental Regulation

National Environmental Policy Act

Chapter 10, which deals with U.S. environmental aspects, describes the need for a National Environmental Policy Act (NEPA) impact statement at various states in the development of an OTEC program. That discussion points toward the probable requirement for a programmatic environmental impact statement which would consider the entire program of OTEC plants at an early stage. Such a requirement would call for scientific studies to provide data for the statement. Because of the potential international environmental aspects of the technology and the placement of OTEC plants in either territorial or international waters, countries other than the United States will be interested in these environmental analyses and their interpretation. Indeed, some other countries may wish to participate in the NEPA process within the United States. There is precedent for the United States circulating NEPA statements to other countries for comments. In 1971, Project Cannikin, the multimegaton underground nuclear explosion on Amchitka Island, Alaska, was of concern to both the Canadian and Japanese governments because of the risk of radioactive contamination, secondary earthquakes, and tidal waves. The impact statement prepared by the Atomic Energy Commission was sent to those governments for comment, and briefings were held by the United States for representatives of those governments who wished to participate.[3] Additionally, in the consideration of the Trans-Alaskan pipeline, a Canadian citizen and a Canadian environmental group were permitted to intervene in the court proceedings held during the NEPA process.[4] The *per curiam* opinion of the U.S. Court of Appeals permitted this intervention on the ground that the interests of the Canadian groups could not be adequately represented by the plaintiffs (U.S. environmental groups). Since several years, perhaps a decade, will elapse before the OTEC program reaches the stage where a program statement will be required, certainly these kinds of procedures will evolve. The Council on Environmental Quality has suggested that:

> [F]ormal procedures might be employed in the future whereby affected countries, especially neighboring nations, would be asked to submit comments that would be circulated and integrated into the NEPA process. Nongovernmental groups in other countries with demonstrated interests might also be asked to participate in the impact statement comment process.[5]

An additional aspect of this problem is the guidelines required by NEPA for agencies engaging in activities which may have a significant effect on the environment.[6] The guidelines are in "final form" but are still subject to further revision in light of evolving administrative and judicial practice. The evolution of the NEPA process may also require inclusion of procedures facilitating participation by foreign countries and nongovernmental organizations. It is understood that the Council on Environmental Quality is considering including additional factors in a further revision of the guidelines.

In a related development, the Council of Ministers of the Organization of Economic Cooperation and Development, in November 1974, approved a series of recommendations on transfrontier pollution. One of these set forth a principle of equal rights of hearing, which is designed to permit the access of interested individuals to administrative and judicial procedures in situations where transfrontier pollution has occurred or may occur.[7]

At later stages in the development of the OTEC program, when specific demonstration plants are being proposed or when the program reaches the stage of commercial operation, similarly NEPA would call for specific impact statements in which other countries might also wish to participate.

If OTEC develops to a commercially viable technology, other countries with appropriate gradients in their territorial or international waters will certainly wish to consider the development of their own sites, possibly through the development of their own technology. Since it is more likely that some of the gradients with energy potential are found off the coasts of developing countries (in the tropical zones for the most part), there may well be a request to the United States for technical assistance or financing of other countries' OTEC operations. It is possible that the evolution of NEPA practice will, in the future, require NEPA statements for activities conducted abroad funded by a U.S. federal agency. This issue has been raised in the United States courts only twice. In the first case, an environmental group filed suit requesting the Export-Import Bank and the AEC to comply with NEPA before they financed or licensed the sale of nuclear parts and materials to another country. The decision in this case was not definitive, and the AEC was requested to file interim procedures dealing with the subject.[8] The EXIM Bank argued that their role was only that of financing and that they were not subject to NEPA since they had no nuclear power program. The court made no determination about their role.[9]

In the second case, the Department of Transportation and the Federal Highway Administration had participated in the funding and construction of a portion of the Pan American Highway in the Darien region of Panama and Colombia. An impact statement had been filed, but the district court held that it was insufficient because it did not adequately treat the health effects (the possibility of the spread of hoof and mouth disease) which could result from the construction of the highway with possible resulting injury in North America. The court said the impact statement also failed to consider the impact of alternatives, and was deficient in that it was not circulated to the Environmental Protection Agency. For these reasons a preliminary injunction was issued, enjoining the FHWA from obligating or expending funds or taking any other action in furtherance of construction of the Darien Gap Highway, until action was taken "to comply fully with the substantive and procedural requirements of NEPA."[10]

By the time the OTEC technology gets to the point that other states are interested in it, probably this problem will be resolved. If the solution is one in which a financing agency is required to examine the environmental effects of its actions–especially if the funds are earmarked for specific projects–then these additional factors will have to be incorporated into the method of operation of the agency.

UN Environmental Programs and International Environmental Law

Although the discussion thus far has been confined to NEPA, it is likely that the evolution of procedures resulting from the 1972 UN Conference on the Human Environment will also have an effect on the need for appropriate notification and consultation with other countries on OTEC. The Stockholm Conference Declaration, building on already accepted principles of international law and practice, developed the basic rule of action that a nation is responsible for environmental harm that it causes, or that is caused by activities carried out under its jurisdiction, where that harm takes place outside its national jurisdiction.[11] The accepted principles and practice are based on a number of precedents, including the Trail Smelter Arbitral decision which stated that no state can use its territory in such a way as to cause injury in the territory of another state[12] and the Corfu Channel case before the International Court of Justice which referred to each state's obligation not to allow its territory knowingly to be used for acts contrary to other states' rights.[13] Moreover, the declaration also contained a principle that:

> States shall take all possible steps to prevent pollution of the Seas by substances that are liable to create hazards to human health, to harm living resources and marine life, to damage amenities or to interfere with other legitimate uses of the Sea.[14]

These principles were carried forward in the discussions to develop a law-of-the-sea treaty, and the Informal Single Negotiating Text (ISNT), developed at the Spring 1975 session, also set forth a description of a nation's responsibilities in this area. Article 2 of the part dealing with protection and preservation of the marine environment provides:

States have the obligation to protect and preserve all the marine environment.[15]

The text was submitted to the conference as a "basis for negotiation," and it does not bind countries. Yet it will be seriously considered, and may well form the framework for an ultimate text, if one should be agreed upon. The ISNT also contains an article setting forth nations' responsibilities to ensure that activities "under their jurisdiction and control" do not cause damage either to the marine environment of other states or to areas of the marine environment beyond the limits of national jurisdiction.[16] An OTEC plant would be a "national" of a particular country, and would, if the convention is approved, be responsible for environmental damage caused internationally.

This discussion has also come up in other contexts at the United Nations. General Assembly Resolution 3129 asked the UN Environment Program (UNEP) to report on measures of establishing "adequate standards for the conservation and harmonious exploitation of natural resources common to two or more States" and for developing "a system of information and prior consultation" among countries sharing such a resource in regard to exploitation.[17] This subject was discussed by the UNEP at the Third Session of the Governing Council in Nairobi, from April 17 to May 2, 1975. A report of the executive director on "Cooperation in the Field of the Environment concerning Natural Resources Shared by Two or More States"[18] was discussed. There was some discussion among UNEP members to limit the consideration of natural resources shared by two or more states to those shared only by a few states, omitting those common to all people. However, adjacent waters facing the open sea or in a semienclosed sea were considered to be within the purview of the discussion.[19] The executive director of UNEP noted that he agreed with this approach.[20]

At this point it is too early in UNEP's investigation to specify with any certainty what the results will be. However, it is not difficult to project certain areas which will be considered. It is likely that UNEP will urge the adoption of a "code of conduct" in this area.[21] Since the oceans are among the greatest of our shared natural resources, it is very possible that a program utilizing the oceans for thermal energy conversion will be considered an activity subject to the code of conduct. Among the elements that may well be included in the code of conduct are a requirement that a country engaging in an activity which may have an environmental impact on a shared natural resource should notify other interested countries of the activity. Moreover, there may well be a requirement for the sharing of information about the environmental impact of the project available to the notifying party. Additionally, states that consider themselves affected may require that there be formal consultation between them and the state launching the OTEC program, to review the effects that it might have. The UNEP document further notes:

> The code should, while recognizing the sovereign right of States to exploit natural resources within their jurisdiction or control, uphold the responsibility of a State to ensure that the exercise of such sovereign right does not cause damage to the environment of other States or of areas beyond the limits of national jurisdiction.[22]

This principle is derived from the principle of responsibility of Stockholm. Finally, it is likely that some form of adjustment-of-dispute mechanism will be developed to consider the questions of disputes raised by the usages of shared natural resources. The law-of-the-sea conference also has before it a draft on dispute settlement which will be discussed at the New York session scheduled for March 1976.[23]

This discussion is necessarily speculative since the development of rules, norms, law, or codes of conduct is still very much in the introductory stage. The outlines and trends are apparent, however, and they point toward an increasing involvement by the international community in activities carried out on or affecting the commons.[24]

Projections for OTEC plants range from a conglomeration of only a few plants to upward of a thousand 400- to 1000-MW plants along the East Coast of the United States. Additionally, some configurations off the shores of other parts of the United States (in the Gulf of Mexico and off Hawaii) also call for collections of these plants. If there are harmful environmental effects, they will increase in number and concentration as the number of plants increases. It certainly is not known when a point will be reached that will push beyond the threshold. The questions raised about the depletion of the ozone layer are analogous. They may or may not be valid claims, but what is important is the risk of ignoring them. For example, there has been some stated concern that a number of these plants along the Gulf Stream could change the temperature and volume of various levels of the stream with possible long-range climatic or weather modification effects. At this point in the development of the OTEC program, whether there will be any effects at all on climate cannot be known, and the assessments are only recently underway. Yet it is one specific area in which there has been some progress by the international community as a result of Stockholm urging a process of notification and consultation. Recommendation 70 of the Stockholm Action Plan stated:

> It is recommended that governments be mindful of activities in which there is an appreciable risk of effects on climate, and to this end: (a) carefully evaluate the likelihood and magnitude of climate effect before embarking on such activities;

(b) consult fully other interested States when activities carrying a risk of such effects are being contemplated or implemented.[25]

UNEP is now in the process of identifying the procedures that can be developed. There have been some proposals put forward by scientific groups with governmental support for a moratorium on activities that may have a large-scale effect on climate.[26] It has further been proposed that a scientific body be established to review the effect of certain proposed activities on climate before those activities are carried out.[27] In one such proposal, where the scientific body finds a risk of harm to the other interested states, the express permission of each of those states to proceed with the action would be required.[28] Further, the same proposal notes that where there are international effects on weather and climate as "the accumulative result of disparate activities," the World Meteorological Organization and the International Council of Scientific Union should establish a panel to assess risks and establish guidelines for this subject.[29]

International Agreements

There are a number of international agreements, either in force or in the process of ratification, which may apply to OTEC plants. A number of these apply specifically to vessels or to vessels and platforms, and it is likely that OTEC plants will be considered one or both of these. Several conventions might apply.

The 1972 Ocean Dumping Convention

This treaty limits the dumping of certain substances from ships.[30] This convention applies to "any deliberate disposal at sea of wastes or other matter from vessels, aircraft, platforms or other manmade structures at sea."[31] "Wastes or matter" are defined as "material and substance of any kind, form or description."[32] In order to regulate this dumping, wastes are divided into three categories: those wastes set forth in annex 1 are prohibited; those set forth in annex 2 require special permit; and all other wastes require a general permit.[33] Permits are issued by domestic agencies in each country.[34] In the United States the Environmental Protection Agency issues the permit. The various substances utilized on an OTEC plant which may be "dumped" will have to be examined to ascertain whether they fall within the restrictions of the convention. If so, recourse will have to be made to the "appropriate authority" of each of the contracting parties to ascertain whether substances fall within the convention. In the United States, the Marine Protection Resource and Sanctuary Act contains the implementing legislation.[35] Implementation and enforcement of the convention are left to each individual party.[36]

The 1973 IMCO Convention for the Prevention of Pollution from Ships

This convention, which is not in force, includes in the definition of *ships* "a vessel of any type whatsoever operating in the marine environment," including "submersibles, floating craft, or fixed and floating platforms."[37] Thus it is clear that an OTEC platform will be considered a ship within the meaning of this convention.[38] Any OTEC platforms, therefore, either flying the flag of the United States or operating under its authority, will be subject to the convention if and when it is ratified by the United States and the requisite number of other countries. The regulations to this convention include annex 2 of the convention which consists of regulations for the control of pollution by noxious liquid substances carried in bulk. This annex is mandatory for all parties to the convention.[39] The annex applies to noxious liquid substances discharged into the sea from tank cleaning or debalasting operations, which would present a major hazard to either marine resources or human health or cause serious harm to amenities, or other legitimate uses of the sea and therefore justifies the application of various types of antipollution measures.[40] Among the substances listed in appendix 2 to that annex is ammonia, which is one of the substances being considered for the working fluid in the OTEC facility. Although propane and Freon are not listed in the annex, it can be amended. The convention provides for amendments to the lists by a qualified majority, i.e., if not opposed by more than one-third of the parties with a combined merchant fleet of not less than 50 percent of gross tonnage.[41] The IMCO convention would apply not only to the carriage of the ammonia to the plant but, since the OTEC facility itself is considered a vessel, to the use of the working fluid while on board the vessel.

Land-based Sources

These sources of pollution include direct discharges into the oceans from shore-based facilities, discharges into rivers which empty into the oceans, and precipitation in the ocean of atmospheric pollution which originates in emissions on land. Although presently there are no international regulations dealing with land-based sources which affect the United States, it is likely that this will be one of the areas of concern for the international community for which broader international agreement will be sought. There is one regional convention, the Paris Convention, which involves fifteen states of the Northeast Atlantic area.[42] If regulations concerning land-based sources of a similar kind are extended to the United States by the conclusion of an international agreement, then both the onshore and offshore activities of OTEC may well be covered. Onshore, the construction, repair, and port facilities may involve pollution of the types described in other similar port facilities which are regulated by United States law.[43] They may well be subject to the international regime, if it is created, established by an international or regional convention dealing with land-based sources of pollution.

Distant Uses of OTEC

The above discussion is concerned primarily with those uses of OTEC which are not distant from shore in that they involve the transmission of electricity back to the coast. It is clear that there may well be other uses of OTEC which could have environmental effects of an international nature for which regulation would be needed. For example, it may be considered desirable to use an OTEC facility to manufacture hydrogen or ammonia or other substances on board which would then be tanked back to shore. The above-mentioned regulations in the 1973 IMCO convention would deal with such transmissions. Additionally, it has been suggested that OTEC facilities may be used to provide energy for the processing of the ores from deep seabed mining operations. The U.S. government is presently investigating some of the potential environmental effects of deep seabed mining.[44] If an OTEC facility is located near such an operation, there is a question as to whether the turbidity raised by the mining of nodules will require a further cleaning of the water before it goes into the OTEC tanks.[45] It is likely that if there is a law-of-the-sea agreement which sets up a regime governing deep seabed mining, then the associated facilities will be included and will be subject to the required environmental regulation. It is not likely that this will work any additional hardship on OTEC that will make it an unattractive source of power for deep seabed mining. Under current plans, however, the first generation of deep-sea mining will probably be processed onshore so that this type of operation is highly speculative. It is considered, however, since it may well provide an attractive source of power for an operation in the deep seabed.

Conclusions

Much of what has been stated is necessarily speculative, a word which often, and with some justification, creeps into discussions of a variety of aspects of the development of OTEC. This discussion has postulated two different uses for OTEC—first as a device to produce energy which will be transmitted to shore, and second as a device to enable activities to take place at sea that, for a variety of reasons, are not desirable onshore.

There will inevitably be environmental effects from these two types of facilities. It cannot be the purpose of this chapter to identify and analyze them all. Rather, what has been attempted is to predict the types of legal mechanisms dealing with environmental aspects—of an international nature—that are relevant and are likely to be adopted in the time projections for OTEC. On the theory that prevention is better than cure, there has been an emphasis on those devices that can assist in anticipating environmental questions. If OTEC receives the blessings of those charged with its technological development, a number of things will happen. First, other countries will become interested in the process and will awaken to both the advantages and the possible harm it can create. Second, there will be a growth in multilateral regulation, including environmental aspects. It is prudent to recognize these factors in the beginning so that adequate planning can be incorporated into the development process. It is also prudent so that the international community can plan to take those measures and adopt those procedures which will be necessary for ensuring that, for OTEC or other ocean-based technologies, effective environmental assessment, and regulatory techniques will be available.

Notes

1. *See* Dept. of Commerce, Deep Ocean Mining Environment Study (April 1975).

2. For preliminary analyses see studies cited in Dugger, ed., *Proceedings, Third Workshop on Ocean Thermal Energy Conversion* (May 8-10, 1975), at 151-179; and Report of a Working Group, Id., at 211, which noted the need for more studies.

3. CEQ *5th Ann. Rep.* (1974) at 400. *See also* Stein, "Cannikin," in Stein, ed., *International Responsibility for Environmental Harm* (in process).

4. Wilderness Society v. Morton, 463 F.2d 1261, 4 ERC 1101 (C.A.D.C. 1972).

5. *5th Ann. Rep., supra* note 3.

6. 38 *Fed. Reg.* 20,550-20,562, August 1, 1973. Also *5th Ann. Rep., supra* note 3, at 506.

7. This principle stated:

Countries should make every effort to introduce, where not already in existence, a system affording equal right of hearing, according to which:

(a) whenever a project, a new activity or a course of conduct may create a significant risk of transfrontier pollution and is investigated by public authorities, those who may be affected by such pollution should have the same rights of standing in judicial or administrative proceedings in the country where it originates as those of that country;

(b) wherever transfrontier pollution gives rise to damage in a country, those who are affected by such pollution should have the same rights or standing in judicial or administrative proceedings in the country where such pollution originates as those of that country, and they should be extended procedural rights equivalent to the rights extended to those of that country.

OECD Doc. C (74) 2124 November 21, 1974.

8. Sierra Club v. A.E.C., 6 ERC 1980 (D.C.D.C. 1974).

9. Id. at 1932.

10. Sierra Club et al. v. Coleman, Civil Action no. 75.1040, (D.C.D.C. Oct. 17, 1975).

11. Stockholm Conference Declaration, Principle no. 21.

12. 3 U.N. Rep. Int'l Arb. Awards 1965 (1941). *See generally* Bramsen, "Transnational Pollution and International Law" at 257, and Stein, "Legal and Institutional Aspects of Transfrontier Pollution" at 285, both in OECD, *Problems in Transfrontier Pollution* (1974).

13. *I.C.J. Reports* (1949) at 22.

14. Stockholm Conference Declaration, Principle no. 7.

15. Informal Single Negotiating Text, A/Conf. 62/WP.8/Part III. Article 2 states: "States have the obligation to protect and preserve all the marine environment."

16. Id., Art. 41.

17. UNGA Res. 3129 (XXVIII). *See also* Chayes and Stein, eds., "The Adjustment and Avoidance of Environmental Disputes: Summary of Discussions of a Conference" [A special publication of the American Society of International Law (1975)].

18. UNEP Doc. GC/44 February 20, 1975.

19. Id. at ¶7.

20. Id. at ¶86.

21. Id. at ¶87.

22. Id.

23. This section of the Informal Single Negotiating Text was submitted by the president of the conference two months after the conclusion of the 1975 session. A/Conf.62/WP 9. July 21, 1975.

24. For example, I understand that recent discussions by the "Evensen group" have resulted in acceptance of international standards for vessels both in the international area and in the economic zone. Since, as is discussed below and in Chapter 6, OTEC plants may be considered "vessels," there may well be construction and discharge standards which would apply.

25. UNEP Action Plan Recommendation no. 70 in the Report of the U.N. Conference on the Human Environment.

26. *Study of Man's Impact on Climate* (SMK) M.I.T. Press (1971); *Study of Man's Impact on Global Environment* (SCEP) M.I.T. Press.

27. *See* "The Adjustment and Avoidance of Environmental Disputes," *supra* note 17, at 21-23.

28. Id.

29. Id. at 23.

30. Convention on the Prevention of Marine Pollution by Dumping of Waste and Other Matter (1972). This convention entered into force on Aug. 30, 1975.

31. Id., Art. III, 1(a).

32. Id., Art. III, 4.

33. Id., Art. IV.

34. Id., Art. VI (1).

35. Pub. L. 92-500.

36. Convention on . . . Dumping, *supra* note 30, Art. VII.

37. Art. 2(4).

38. For a similar conclusion, *see* Chapter 9.

39. Certain annexes (3, 4, and 5) are designated as optional by Article 14 of the convention.

40. Annex II, Reg. 3(1)(a)-(d).

41. Art. 15 of the convention.

42. Concluded on February 19, 1974; not yet in force.

43. *See* Chapters 9 and 10.

44. *See* the Report of the Department of Commerce in DOMES I.

45. *See* discussion in study cited *supra* note 1. *See also* Frank, "Deep Sea Mining and the Environment," a report prepared for the ASIL (Nov. 1975).

8 Problems of Legal Responsibility and Liability to Be Anticipated in OTEC Operations

J. Daniel Nyhart

Introduction: Nature and Sources of Liability, Parties

The concern of this chapter is the most significant legal liabilities likely to accrue to the major parties if an OTEC operation becomes reality. The legal arrangements for handling them are also discussed. A recurring question is whether an OTEC plant,[1] located in and on the ocean, using elements of the ocean—heat, cold, fluidity, space—as essential and integral parts of its functioning, would best be assimilated for such legal purposes into the body of maritime law, which had its origin in, and owes many of its characteristics to, the shipping industry's needs for a legal framework. Or would some format for extending "land law" be a better solution? Or perhaps would some new combination be best? In these respects, OTEC is prototypical of many new uses of the ocean which may assume a place alongside such traditional uses as transportation and fishing.

Nature of Responsibility and Liability

The advent of demonstration OTEC plants off U.S. coastal waters (a possibility as early as 1985) and later off selected tropical sites would create numerous responsibilities to adhere to legal standards of care and conduct, as well as liabilities for failure to do so. The responsibilities of concern are those obligations articulated largely as standards applicable to, or as principles governing, agreements among the extensive variety of parties who would be involved in an OTEC operation. *Liabilities* refer to legal sanctions, most frequently some form of deprivation, attaching to one or more parties for failure to meet a standard or to follow a principle of conduct concerning an agreement or accepted legal norm.

The responsibilities and liabilities include those arising from ownership and risk of ultimate loss, construction, operational functioning, navigational hazards and collision, environmental considerations, injury and wrongful death, and maintenance of public order and welfare.

Many parties are likely to be involved. They encompass investors and owners, who may comprise a complex of mortgagors, lessees, trustees, front entities, and operating interests. They also include operating interests, those with day-to-day responsibility arising either contractually or from their ownership of the plant. Personnel on the OTEC structure, who may work for one or more operating or service entities, will have important rights as well as responsibilities. Technical designers and engineers who design and build the plant, its housing, and positioning elements may be parties to some liability issues. And so will vessels tying up to the OTEC structure, and those on them. There are also strangers, not directly involved, who may collide with an OTEC operation, damage or destroy it, steal parts of it, or possibly salvage it. Governmental entities will regulate, finance, or own an interest in an OTEC plant. And, finally, classification societies and insurers, who set pertinent standards and evaluate risks, will also play a part.

Assumption as to Geographical Jurisdiction

The OTEC operations foreseen as prototype plants[2] in this chapter would be operating outside the presently existing boundaries of the U.S. territorial sea and the contiguous zone and therefore currently would be beyond U.S. geographical jurisdiction. The assumption is made here that the present Outer Continental Shelf Lands Act[3] does not offer jurisdictional color, because there is no direct activity involving exploration or exploitation of the seabed or subsoil. Thus OTEC is one of a growing number of potential uses of the ocean—multiple-use platforms, aquaculture, water-column chemical extraction, etc.—which would have, excepting admiralty, no evident jurisdictional basis beyond existing territorial waters, even though over the continental shelf.

However, it is probable that by 1985—a possible early date contemplated for U.S. prototype OTEC operation—the United States will be exercising regulatory jurisdiction out to 200 mi for purposes of economic exploitation of the surface and water column. Chapter 3 contains the background leading to this conclusion that, either through international treaty agreement or by extension unilaterally, the United States will have an economic zone. A unilateral extension might be made either by a blanket legislative enactment or by a series of acts, each aimed at a specific use of the oceans.

Hence, it is assumed that a floating, moored OTEC plant operating within 200 mi off the U.S. coasts (thus over the U.S. continental shelf) would be brought within U.S. jurisdictional reach through legislation extending jurisdiction and the applicability of federal laws generally to new economic uses of the surface and water column in addition to those of the seabed and subsurface. Alternatively, there might be specific OTEC legislation with similar goals.

Sources of Liability

There are five potential bodies of law from which liability could arise for failure to meet standards pertaining to an offshore OTEC plant. The first three are federal law.

Federal Regulatory Law. First, there is the federal statutory and administrative "regulatory" law, aimed, in the broadest sense, at furthering public safety, health, or welfare objectives of the nation—in short, those activities falling under the police power as it exists at the federal level (see Chapter 9). They cover construction, layout and design, financing, initial certification and periodic inspections, at-sea and operational safety, customs, and navigational and environmental controls.[4] Chapter 11 makes reference to laws pertaining to financial and corporate structure. Chapters 7 and 10 identify pertinent environmental regulations. Many statutes or regulations specifically state the sanctions to which those failing to meet the related standards will be liable. In addition, it can be argued that the existence of standards, and of conduct falling below those standards, lends evidentiary weight under a civil proceeding by one party seeking to pin liability on another. Liability aspects of the various U.S. federal regulatory regimes will be brought into the discussion below as applicable.

U.S. Civil and Criminal Law. Other issues in this chapter deal with responsibilities and liabilities arising from long-standing principles of tort, contract, property, and personal injury law, which have found their way into the federal judicial system through invocation of federal question or diversity routes.[5] Obviously, decisional law is supplemented by statutes, which in turn are interpreted by judges. And some of the statutory supplementation can be classified as "regulatory" and so could belong in the paragraph above. The important points are that there is a body of legal principles aimed at resolving tort, contract, and similar disputes in the federal courts; that the jurisdiction of the courts will be extended to cover an offshore OTEC operation; and that the civil and criminal laws of the United States are likely to apply.[6]

U.S. Maritime Law. Admiralty (or maritime) law, while it may be considered a subset of U.S. law, offers a distinct choice of applicable law for many OTEC purposes, provided they fall under admiralty jurisdiction. Admiralty law has been defined as:

> [A] corpus of rules, concepts, and legal practices governing certain centrally important concerns of the business of carrying goods and passengers by water. . . . The subject comprises the most important part of the private law that deals with the shipping industry. . . . in some modern cases it has even been held to cover some matters quite unconnected with shipping although the Supreme Court has recently cut back quite sharply on this development.[7]

Maritime law "arose and exists to deal with problems that call for legal solution, arising out of the conduct of the sea transport industry."[8]

Substantively, maritime law in the United States comprises two parts: the "general maritime law"—the body of traditional rules and concepts long associated with navigation and the shipping industry, modified by rules and concepts developed by the judiciary specifically to fit the needs of this country; and the body of statutory law that Congress has passed to supplement or alter the first part.[9]

Since 1966, admiralty cases in U.S. courts have been dealt with under the same docket, procedures (with some exceptions), and nomenclature as other civil cases in federal courts. However, it yet provides a separate ground for jurisdiction, and a person bringing a suit may invoke this ground in his or her complaint, thereby bringing into play any special rules for handling admiralty law.[10]

Thus U.S. maritime law, if it applies, provides its own jurisdictional reach, has some distinctive procedural aspects which have carried over under the 1966 unification, and has a distinct body of substantive law, about which more will be said.

International Law. The international regulatory agreements reviewed in Chapter 6 also will impose liability to those involved in OTEC, although most probably it will be implemented through U.S. legislative enactments of the type referred to in the prior paragraphs. Chapter 3 reviews the prospects for extension of international sources. The liability side of all these, too, will be brought into focus in the appropriate places.

State Law. In the United States, the basic common law of tort, contract, property, etc., lies in state law. In addition, many coastal states have laws pertaining to "maritime" matters or more general subjects, such as environmental law, which affect uses of the oceans.

The legal framework for handling issues of liability arising from OTEC operations would most likely include all these sources, whether by congressional design or evolution of existing law applied to cases as they arise. It is a reasonably safe assumption that, like other significant technological innovations, if developed, OTEC will evoke a legislative/regulatory response.[11] Such a legislative initiative ought to comprehend a means, or set a direction, for handling the issues of liability referred to in this chapter.

There are two major questions: When ought this particular use of the oceans be brought under maritime law, other U.S. law, or state law, for purposes of establishing and handling liabilities? How should the problem be handled by Congress when it addresses itself to the task of legislating for OTEC operations? These questions are discussed in the section Approaches to the Problem of Legislation regarding OTEC Liability as to substantive law, and in the section A Forum for OTEC Liability Issues. The section Liability Issues provides illustrations of the kinds of issues that may arise. First, the applicability of the major existing body of law pertinent to these issues is examined in the following section.

Applicability of Maritime Law to Nontraditional Uses of the Ocean

In order to understand the existing legal environment which will surround OTEC if it comes into existence, it is essential to examine its relationship to admiralty law.

With regard to liabilities and responsibilities, maritime (or admiralty) law, chief among the spectrum of sources identified above, in the past has provided the framework for handling the issues and cases arising from use of the oceans. But whether it will or ought to continue to do so as new technological uses of the ocean develop is an open question. Future policy determinations are likely to suggest maritime law solutions for some purposes and not for others.

The Nature of Admiralty Jurisdiction

Admiralty is one key into the federal courts. The Constitution[12] and the Judiciary Act of 1789[13] gave federal district courts exclusive original jurisdiction over maritime cases, "saving to suitors, in all cases, the right of a common law remedy, where the common law is competent to give it."[14] Admiralty may offer the only constitutional and statutory grounds for jurisdiction. If so, jurisdiction lies independently of diversity, federal questions, or amount in controversy.[15]

In admiralty law, it is possible for one to sue the vessel involved in an *in rem* proceeding, in addition to seeking satisfaction from individuals or corporate entities. The ability to gain a maritime lien against the vessel in satisfaction of a judgment is a closely related characteristic. Vessel owners may, under U.S. statutes, in certain conditions limit their liability to the value of the vessel. In maritime law, equitable remedies and jury trials are not available; nor are statutes of limitations normally imposed. The principles for arriving at liability determinations frequently differ from regular tort remedies.

These are only some of the differences arising from a determination of whether an issue is a maritime one.

Today, the scope of the jurisdiction is fuzzy at the edges, right at the place an OTEC operation and other new technologies would find themselves. The existing law on the extent of maritime jurisdiction is comprised of a multitude of statutes and court cases dealing with numerous specific fields—personal injury issues, maritime lien issues, insurance issues, etc. Each has to be taken separately. It is unlikely that a uniform answer of maritime or not-maritime would apply today to the range of OTEC liability issues.

Story wrote early that admiralty

> . . . comprehends all maritime contracts, torts and injuries. The latter branch is necessarily bounded by locality; the former extends over all contracts . . . which relate to the navigation, business or commerce of the sea.[16]

For contract cases this relationship has been dispositive.

But for tort purposes, "maritime" character has been further defined in terms of three possible qualifications.[17]

First, the jurisdiction exists for events relating to water, within and beyond territorial waters, reaching onto the high seas, and including foreign navigable waters.[18] Most operations related to OTEC would meet this qualification.

Second, for many purposes maritime jurisdiction deals "only with vessels and their cargos and personnel."[19] This requirement—that the involved structure qualify as a vessel—is pervasive. For many courts, for example, it is the first question among many involved in determining liability for injury or death. Gilmore and Black call it "the first fundamental test of admiralty jurisdiction."[20] The question reaches beyond the context of personal injury and death. It is controlling or significant in many statutes which have bearing on liability issues—financing (the Federal Maritime Lien Act[21] and the Ship Mortgage Act[22]), flag (documentation system),[23] and limitation of liability (Limited Liability Act).[24]

The third possible qualification in bringing a tort issue within maritime law is that the wrong must bear a significant relationship to traditional maritime activities. Maritime locality alone no longer appears sufficient. This requirement has been enunciated recently by the Supreme Court in an aviation tort case,[25] but soon it may well extend beyond that class.

Until that case, whether or not an OTEC structure would be characterized as a vessel would probably have been dispositive of the question of maritime law applicability in most liability issues. If it were considered a vessel, admiralty law would apply in the absence of controlling legislation. However, in making the new requirement,[26] the Court raised broad policy considerations as to the desirability of broadening the application of maritime law. The case's meaning for new technology must be examined.

OTEC as a Vessel

It is highly probable that under existing statutes, regulations, and case law, an OTEC structure would, or certainly could, be classified as a vessel for most purposes pertaining to liability issues. An early statutory definition states:

> The word "vessel" includes every description of watercraft or other artificial contrivance used, or capable of being used, as a means of transportation on water.[27]

This definition has served as a basis for general subsequent statutory enactments. The language, "used, or capable of being used as a means of transportation on water," on the one hand, and Story's phraseology, "which relate to the navigation, business or commerce of the sea," on the other, mark the parameters within which most court decisions have been made as to what constitutes a vessel. They also furnish the key words—*transportation, navigation* and *commerce.*

All sorts of structures and craft normally not coming to mind as a ship have been classified as "vessels" and therefore brought under general maritime law and particular maritime statutory enactments.

Barges,[28] dredges,[29] derrick hoists,[30] pile drivers,[31] mobile oil drilling and production rigs,[32] loading elevators[33] and cranes,[34] pumpboats,[35] target boats[36] and scows,[37] moth-balled landing craft[38]—all craft with no propulsion of their own and which consequently have to be towed—have been classified as vessels for purposes, variously, of general maritime jurisdiction, maritime liens,[39] limitation-of-liability issues,[40] personal injury and death cases,[41] and federal inspection purposes.[42]

Thus the fact that an OTEC structure may not have propulsion and would have to be towed into operating position does not appear to be a bar under decisional law to its being classified as a vessel for purposes of maritime law.

A second category of structures, those moored more or less permanently to the shore or pier—such as a dancing platform or barge,[43] restaurant,[44] houseboat,[45] refrigerated cargo carrier converted into a refrigerated shrimp processing plant[46]—have also been found to be vessels.[47] Thus the prospect of OTEC plants being moored and/or connected to the beach by power-transmitting cable would not seem to be a block to the applicability of maritime law—at least for many purposes.

The rationale underlying these decisions varies. Use of the structure for transportation purposes includes such nonshipping functions as carriage of sand and gravel—and therefore encompasses many types of dredges.[48] But transportation also has meant carriage of a permanent cargo (e.g., engines and other equipment to do particular work); hence a pumpboat,[49] which carries neither passengers nor freight, is brought into the jurisdiction. *Navigation* has been taken to mean capable of being navigated, even though very inefficiently, with the structure built for a purpose requiring navigation, and not necessarily involved in navigation at the time under consideration.

Where statutory definition has been explicitly relied on as being integral to the issue,[50] decisions have used the "capable of being used" language broadly. For example, in the field of maritime liens for supplies and repairs,

> [I]t is now generally recognized that the statutory definition of "vessel" is basic to the decision of any particular facts, and that definition is set forth under the general provisions of Title 1, United States Code in §3.[51]

In 1950, thus, a navy bomb target boat was held within the statute and "whether [it] . . . had ever sailed the seas or transported any portion of the world's commerce [was] therefore not controlling."[52]

The philosophy underlying the extension of admiralty to special-purpose craft is summarized by an 1896 case dealing with maritime liens for seamen's wages:

> The idea of commerce does not come into the mind primarily in connection with such craft; but, when it is borne in mind that they are constructed to move upon the water, and nowhere else, and that, while thus moving upon the water, they are subject to all the rules that govern other water craft as to lights, collisions, etc., it will be seen that they have that mobility and capacity to navigate which are recognized as the prime elements in determining the subjects of maritime [liens for seamen's wages].[53]

However, amid the cases applying maritime law, there has also run a minority-decision thread which reads the application of maritime jurisdiction more narrowly. First, there have been certain kinds of structures which generally have been held not to be vessels. Dry docks,[54] artificial islands and structures for oil and gas drilling and exploitation,[55] and wharfboats[56] generally have been held not to be vessels or to be outside admiralty jurisdiction. The mechanistic approach of comparing a new kind of structure (construction platform) to an old one (floating dry dock) recently was instrumental in restricting the application of maritime law in an injury case on a construction barge tied to a dock in a concrete yard.[57]

For the most part, the cases denying vessel status emphasized connection to existing navigability and commerce. They argue that admiralty properly deals with vessels that "plough the seas," which are engaged in maritime employment, and with contracts touching navigation.[58] Thus, as early as 1897 some courts held dredges beyond admiralty because deepening navigable waters was not maritime employment, which must be related to trade and commerce upon navigable waters.[59] Other cases have banned vessel status from structures not at the time engaged in a "maritime venture"[60] or "maritime commerce or navigation."[61]

Overall, however, the weight of existing decisional opinion, statutory or regulatory direction, and experience provide ample basis for considering an OTEC structure of the future as a vessel and thus within general maritime jurisdiction. The position is probably best summarized in the current context of oil platform personal-injury cases.

The Outer Continental Shelf Era

The exploration and exploitation of oil and gas on the continental shelf offshore the United States (and subsequently elsewhere) was the first major technological advance into the oceans which added a new use alongside transportation and fishing. Therefore it raised substantial problems concerning the handling of liability issues. Prior questions concerning the assimilation of a new technology into the maritime law framework were minor when compared to those of oil and gas.

Basically, three major events mark the liability aspects of oil and gas development of the past 20 years: (1) the lower courts continued to expand the application of maritime law to most types of structures used in oil and gas exploration and drilling, classifying them as vessels; (2) the Coast Guard, which is the administrative agency most closely concerned with regulatory issues, accommodated the new technology into its regulatory framework which is basically oriented around vessels; and (3) the Supreme Court, fairly late in the game, withdrew a major type of structure (artificial islands and structures) from maritime jurisdiction for a major issue area (personal injury and death) and in the process raised questions about the status of offshore structures generally.[62]

Lower Court Application of Maritime Law. In 1959 the Fifth Circuit held that for purposes of general maritime jurisdiction and the Jones Act (dealing with personal injury and death remedies), a mobile drilling platform or barge, moved by tugs into position and resting on the bottom of the Gulf, being jacked out of the water on its own legs, was a vessel.[63] A distinction was made between such special-purpose vessels and a manmade island.

Since then that circuit has continued to apply maritime law to offshore structures engaged in oil and gas exploration and exploitation, with the exception of permanently fixed platforms, discussed below in the subsection Supreme Court Curbing of Maritime Law Application.

Four years later, a quarterboat used for housing employees engaged in offshore oil development was placed within the vessel category by the Federal District Court in Louisiana.[64] The Court cited the statutory definition found in 1 U.S.C. 3, discussed above.

In 1966, a submersible drilling barge stabilized in navigable waters[65] and another resting on the bottom of a canal[66] were both considered vessels for purposes of general maritime law and the Jones Act.

In 1975, a submerged oil-storage facility resting on the bottom in the Gulf of Mexico was held a vessel, again within the provisions of the Jones Act and general maritime jurisdiction.[67] The case is noteworthy in that it was written six years after the Supreme Court held that maritime law was not to be applied to personal injury and death cases on artificial islands and fixed structures. (See discussion of *Rodrigue v. Aetna Casualty and Surety Co.*, below.)

The context of these lower-court cases is personal injury and death, and their application of maritime law has in part reflected the nationwide trend to providing broader recovery to a broader category of plaintiffs. But the typical phraseology of the courts' jurisdictional language includes "for purposes of general maritime jurisdiction. . . ." Thus although admiralty's application has arisen in this proplaintiff milieu, it will influence the issue of applicability in other fields, such as environment, personnel relations, or limitations of liability.

The district court's instruction to the jury in the 1975 case made room for nearly all forms of structures to be included as vessels, except for those permanently affixed to the seabed, that is, those covered by Rodrigue. The instruction may summarize the position of the Fifth Circuit.[68]

Administrative Accommodation. Meanwhile, the Coast Guard, charged with considerable regulatory responsibility on the outer continental shelf (see Chapter 8) has striven to cope with the classification of the new structures in its regulations and administrative practice.

The Coast Guard has created a regulatory definition of an *industrial vessel* which reflects decisional law and which includes Section 90.10-16, 46 *Code of Federal Regulations:*

> [E] very vessel which by reason of its special outfit, purpose, design, or function engages in certain industrial ventures. Included in this classification are such vessels as drill rigs, missile range ships, dredges, cable layers, derrick barges, pipe laying barges, construction and wrecking barges. Excluded from this classification are vessels carrying freight for hire or engaged in oceanography, limnology, or the fishing industry.

This regulation is as yet undeveloped. That is, its impact and meaning have not yet been evolved elsewhere in the regulations. Nor apparently is there yet judicial interpretation of the term. It grew largely out of the rapid technological developments in offshore oil and gas drilling platforms during the last two decades.

In administrative practice, floating (e.g., ship), configured, semisubmersible, submersible but floating platforms, and "jack-up" rigs whenever they are not jacked up are all considered vessels by the Coast Guard for inspection, certification, and navigation purposes. Thus all oil and gas industry platforms other than those permanently affixed to the seabed are administratively regarded as industrial vessels, at least during construction and while afloat. Those permanently affixed are considered *artificial islands or structures.* The term derives from the language of the Outer Continental Shelf Lands Act (§ § 1333 *et seq.*) but is not defined there in terms of the variety of structures now employed in offshore drilling and production. However, in 33 C.R.F. § 140.10-5, the Coast Guard has defined the term to mean:

> [A] building or platform secured to the seabed by fixed means or submerged onto the seabed so that for practical purposes it becomes stationary. This includes both mobile and built-up platforms.[69]

These administrative definitions and accompanying practice provide a basis for concluding that, under existing regulatory law, most likely configurations of an OTEC structure would be classified as a vessel for most regulatory purposes. A dynamically positioned OTEC structure would fit. And a moored OTEC device swinging at the end of a single mooring system would seem more nearly an industrial vessel than an artificial island or structure, on the basis of these regulatory definitions.[70]

Supreme Court Curbing of Maritime Law Application. In 1969 the Supreme Court held that admiralty had no jurisdiction for wrongful death actions under the Death on the High Seas Act, which arose out of accidents on artificial-island drilling rigs beyond the state's territorial jurisdiction and were therefore governed by the Outer Continental Shelf Lands Act (OCSLA). The decision came amid the continued expansion of the application of maritime law to unusual or specialized offshore structures categorized as special-purpose craft or industrial vessels.

Referring to the "artificial islands and structures" named in OCSLA, the Court, in *Rodrigue v. Aetna Casualty Insurance Company*, 395 U.S. 352, using language which could be taken to apply to more than just permanently affixed structures on the continental shelf, said that

> [The] accidents had no more connection with the ordinary stuff of admiralty than do accidents on piers.[71]

Further, Congress's approach of treating the platforms as federal enclaves was

> deliberately taken in lieu of treating the structures as vessels, to which admiralty law supplemented by the law of the jurisdiction of the vessel's owner would apply.[72]

The Court referred to a speech of the senator introducing the OCSLA, stating that the drafting committee had initially considered, and then rejected, the application of admiralty law since it was "not an adequate and complete answer to the problem."

The fact that workers on the offshore platforms were closely connected with onshore facilities and their homes, and frequently commuted to the platforms, etc., led the Court to make the adjacent state law applicable. Finally, the Court was concerned that if maritime law were adopted, foreign flag jurisdiction considerations would control.

Since Rodrigue, whose meaning was expanded upon at least once by the Court,[73] the lower courts have treated the case with circumspection. The Fifth Circuit, for example, at times has followed it in a straight manner, applying adjacent state law to injuries and deaths on fixed platforms.[74] And a Louisiana federal district court has applied the Rodrigue principle to the interpretation of a contract related to an injury case, saying that since the cause of action was governed by state law under Rodrigue, "*a fortiori*, such law applied to interpretation of the contract between contractor and sub-contractor containing an indemnity clause."[75]

As noted, however, the lower courts have continued to apply federal maritime law to other kinds of platforms that were not affixed to the seabed. The situation has led to the anomaly of having fixed platforms surrounded on all sides by various movable platforms with liability for injuries and death being treated quite differently, though the work and hazards to workers are nearly identical.

The Fifth Circuit has also held that the federal Longshoremen and Harbor Workers Act is a total remedy in a Rodrigue situation and therefore refused to apply the Louisiana state compensation law, saying that there is no gap for a Rodrigue surrogate application in this statutory compensation area.[76] This, too, may amount to a frustration of the Supreme Court's objective.[77] The circuit has also held that Rodrigue did not require the application of state law for the damage caused by a collision of a vessel with a platform.[78]

In California, the Ninth Circuit, in a case involving pleasure boats, felt obliged to set out its view on limits to Rodrigue by saying that the OCSLA "does not preclude the application of maritime law to claims with the maritime nexus wholly apart from location of fixed platform on navigable waters."[79]

Potentially, the distinction between industrial vessels and artificial islands and structures holds considerable significance for OTEC in light of Rodrigue, for although the issues in Rodrigue are couched in terms of personal injury or death statutes and maritime law policy considerations, rather than whether a fixed platform is a vessel, the case has philosophical antecedents in the earlier minority group of cases denying vessel status to certain structures.

Is OTEC's Maritime Nexus an Issue?–Rodrigue and Executive Jet Aviation

Rodrigue may have been an early signal of a general holding the line on expansion of maritime law in tort cases.[80] Having seen lower courts include within admiralty surfboard accidents[81] and motorboat accidents on inland lakes,[82] the Supreme Court recently indicated that the question of maritime connection over and above location on water may be increasingly asked during the period in which OTEC will be developed.

In *Executive Jet Aviation, Inc. v. City of Cleveland* (Ohio) (93 S. Ct. 493, 1972), the Supreme Court refused to apply maritime law in the case of a land-based aircraft on an overland flight which struck a flock of seagulls over the runway after a takeoff from a Cleveland airport. The airplane fell into and sank in the navigable waters of Lake Erie, a short distance offshore. The Court said that "maritime locality alone is not a sufficient predicate for admiralty jurisdiction in aviation tort cases." "The mere fact that the alleged wrong occurs" or "is located" on or over navigable waters is not of itself sufficient to turn a case of airplane negligence into "a maritime tort." In this case the Court found there was not significant relationship between such event and "traditional maritime activity involving navigation and commerce." It reviewed the federal system's adherence to a strict locality basis and noted that a mechanical application of that rule, despite the lack of any connection between the wrong and "traditional forms of maritime commerce and navigation," has been sufficient to bring into action the "full panoply of substantive admiralty law."[83] The Court also found some history of requiring that a traditional maritime activity as well as locality is necessary to invoke admiralty jurisdiction over torts.

Summarizing its review, the Court said that "in determining whether there is admiralty jurisdiction over a particular tort or class of torts, reliance on the relationship of the wrong to traditional maritime activity is often more sensible and more consonant with the purposes of maritime law than is a purely mechanical application of the locality test."[84] A maritime nexus similar to that in contract issues is needed.

Executive Jet raises in modern-day form the question asked in the old "vessel" cases confining admiralty's reach to traditional maritime transportation and commerce: What category of activities should fall within maritime law? What is encompassed within traditional maritime activities?

The problem, of course, is that OTEC and other potential new ocean uses are not traditional in the above sense. Moreover, thermal gradient-differential energy production is a technology rooted in the ocean, not merely transplanted there from land. It involves the use of the oceans as an essential part of its operation. Thus Executive Jet's question is not OTEC's. OTEC has a maritime nexus, albeit not a traditional one.

Some legal framework to handle liability issues and other matters is going to be necessary. Congress or the courts are going to have to decide what that law will be: maritime law, state law, or a combination. It was asserted long ago that maritime law has the capacity to accommodate technological change:

> [S]o far-reaching are the principles which underlie the jurisdiction of the courts of admiralty that they adapt themselves to all the new kinds of property and new sets of operatives and new conditions which are brought into existence in the progress of the world.[85]

The question a prospective OTEC development posits is how a new technology is to be assimilated into the legal environment.

A Forum for OTEC Liability Issues

Involvement of the various parties introduced earlier will give rise to a mass of liability claims whose existence is predictable but whose specific nature is not. Accidents will cause injuries and death; workers will become sick, and their sickness may be work-related; disputes will cause fights; supplies or products will be damaged; components will fail; operations will pollute surrounding waters—all in the course of ongoing operations. Catastrophic loss of the whole structure would give rise to an additional variety of claims.

How such cases would be handled begins with two basic questions. What forum would have jurisdiction? What would be the appropriate applicable body of law? In light of the foregoing arguments that U.S. jurisdiction will be extended seaward, that new legislation is likely, and that admiralty law is a body of law pertinent to future OTEC operations, it is useful at this point to examine the issue of forum.[86]

If the United States passed no new laws and if the OTEC structure were operating beyond the territorial seas, the existence of a forum would be dependent upon the claim being recognized as a maritime tort falling under admiralty jurisdiction. Personal injury and death claims to seamen and to others aboard a vessel, a case for a death action alleging that defects in ship design ultimately lead to the death, damage claims resulting from oil pollution impacting a shore installation, and, of course, damage to ships or cargoes on navigable waters and suits for damages arising from collision—all these fall into the category of maritime tort.

Granted the accuracy of the analysis in the previous section, the U.S. federal district courts would have jurisdiction, assuming plaintiffs with standing and appropriate service of defendants.[87]

If the OTEC plant were operating within U.S. territorial waters, maritime law would also apply, and hence U.S. district courts would have jurisdiction too. Also, if the suit were one *in personam*, and if there were diversity of citizenship and required jurisdictional amount, the plaintiff would have an option of bringing her or his case on the "civil" side of the "unified" system under the "saving to suitors" clause.[88] State courts would also be available under the saving-to-suitors clause in an ordinary civil action.[89] In the latter two instances, jury trials probably would be available.

But the United States, along with other nations, probably will not leave ocean jurisdictional issues to the fortunes of present law. More likely, OTEC and other offshore structures and uses will be brought within U.S. political and legal jurisdiction either under legislation implementing an international law-of-the-sea agreement or as part of a general unilateral extension of jurisdiction over an economic zone, reaching beyond present limitations of exploration and exploitation of the continental shelf (see Chapters 3 and 4). Or there might be new legislation extending jurisdiction over OTEC plants specifically within an economic zone or its equivalent. Finally, such legislation might assert jurisdiction over U.S. "flag" OTEC plants on a global basis wherever they might be operating.

Past experience suggests that U.S. federal district courts would be given original jurisdiction over cases and controversies arising out of OTEC operations and related activities. In the prior two analogous instances of assertion of effective jurisdiction beyond the territorial seas, i.e., the Outer Continental Shelf Lands Act[90] and the Deepwater Port Act of 1974,[91] the district courts were given jurisdiction (see the section Approaches to the Problem of Legislation Regarding OTEC Liability).

To the extent new U.S. legislation limits asserted jurisdiction geographically to an economic zone or some equivalent, the maritime jurisdictional reach presumably would continue to provide a forum in U.S. courts for cases arising on the high seas beyond those limits for appropriate maritime torts.

Finally, there is the additional possibility[92] that an international tribunal established under the law-of-the-sea conference might have jurisdiction over cases arising on the high seas or conceivably under certain circumstances within coastal states' economic zones.

Thus it seems likely that a forum will exist to hear liability cases arising from OTEC operations. But how the substantive law will treat such cases is another matter—one that turns largely on what body or bodies of law are deemed applicable. This question is examined in the section Approaches to the Problem of Legislation regarding OTEC Liability. Several issues pertinent to potentially significant areas of liability are examined in the next section.

Liability Issues

The liability issues which would be encountered in the lifetime of an OTEC structure might well range over the whole of tort and maritime law. This section poses some questions which future policymakers will have to answer or leave by default to the courts as actual cases arise. Two areas—liability of owners generally and liability for personal injury and death—are examined in more detail than the remainder.

Liability of Owners

Leveraged leasing or ship-mortgage financing are the two most likely forms of financing for commercial or perhaps even prototype OTEC operation, omitting the possibility of the federal government's funding through an equity or direct lending route (see Chapter 11). In either leveraged leasing or ship-mortgage financing, there appear to be important questions concerning the liability of the owners of OTEC, and even of their identity. The principal reason is that in either form of financing the party operating the OTEC plant would formally be a lessee of the structure, whereas "ownership" would reside in one or more parties who had little or nothing to do with the operation.

This format would be an adaptation of current models of oil platform and large ship financings. Leasing, leveraged or not, is also a current financial tool in many other large capital financings. Thus some of the questions raise issues of creditors' rights which would arise as to liability in any leasing situation. But most arise from the maritime nexus, that is, the degree to which financing patterns derived from shipping and rig financing would carry with them maritime law appurtenances. The lease would almost certainly be written in "land" terminology, eschewing the maritime terms of *charter party, charterer*, etc. Yet the degree to which offshore structures are considered vessels in other contexts has already been explored.

Figures 8-1 and 8-2 outline the relationships giving rise to the questions.

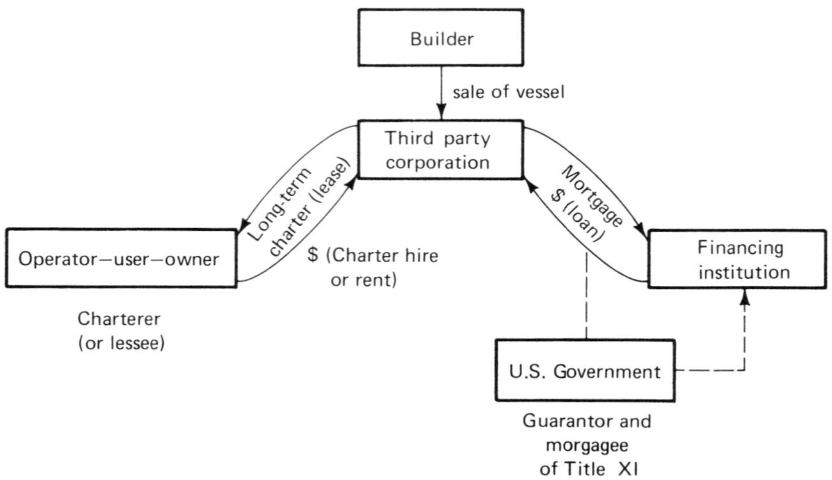

Figure 8-1. Ship Mortgaging.

Leveraged leasing platform financing today, as described in Chapter 11 and as practiced, can be explained essentially as an elaboration of this three-way pattern. Title to the platform is held by a bank as trustee for other financial institutions who own the equity interest in the platform and who have put up a portion (say 20 percent) of the funding through equity. The owner trustee leases the platform to the operating company, whose obligations are guaranteed by its parent corporation. The owner trustee, however, assigns all its interests in the lease, guarantee, and rent to a third-party corporation (created for the purpose) to which it sells a trust note. The funds from this sale plus the equity investment pay the builders for the platform. The third-party corporation raises the money to buy the trust note by issuing its own equipment notes, which in turn are purchased by other financial institutions. These institutions are the true lenders of the 80 percent of the cost of equipment remaining over and above the 20 percent equity. It is likely that they may be identical to, or affiliated with, the equity owner institutions. As security, the third-party institution assigns to a lender trustee the interest it was assigned by the owner trustee.

In terms of Figure 8-1, the third-party corporation has become a third-party corporation plus an owners' trustee (the legal "owner") plus a group of owners formed into a trust. The financial institution has become many, with a trustee holding the rights which collectively form the security on their loans.

By these means, the tax benefits identified in Chapter 11 flow to the financial institutions putting up the funds. One additional change is that,

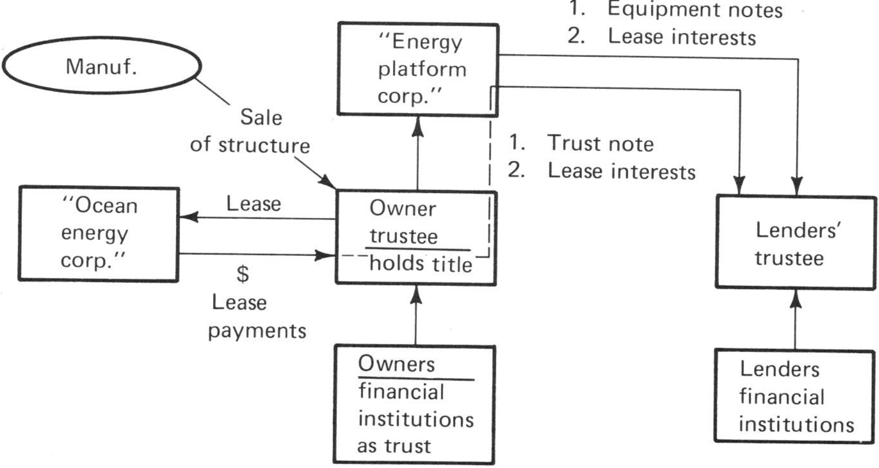

Figure 8-2. Leveraged Leasing.

apparently, the operator-owner-user has not put up any of the equity and does not control the third-party corporation—a change from traditional ship-mortgage financing. (And the equipment notes have replaced the mortgage.)

If this pattern were adopted for financing commercial stage OTEC structures, the situation might look something like Figure 8-2.[93]

The major liability questions raised by this pattern would be: Who bears what liability in case of a disaster in which the structure was lost, along with the lives of those on board, and in which harm was caused to the surrounding environment and perhaps to other structures, such as a colliding vessel?

The intent among the parties presumably would be that the lessee—the operator-user—would bear the liability. Today's platform financing agreements provide that the lessee indemnify the owner trustee, owners, third-party corporation, lenders, and lender trustee from all acts and liabilities arising from the operation and use of the platform.

But presumably outside claimants would not be so bound. The Ocean Energy Corporation might be a very thin operating shell with little or no equity cushion in the operation and no equity interest in the structure. The losses probably would, but might not, be covered by insurance. The operating company might be thrown into insolvency by the loss of the structure. Claimants under those who lost their lives, the state whose shores are damaged, the fishermen whose fisheries were damaged by loss of the heat-exchanger working fluid, and the owner of the vessel in collision with the OTEC—all might understandably seek to go after the owner trustee who holds title, or the owner financial institutions.

However, the claimants may be frustrated in their attempts to reach some "owners" of substance. The owner trustee, which holds title to the OTEC plant, has assigned all its interests in the lease—the principle other asset in the trust estate—to Energy Platform Corporation, the third-party shell, which in turn has assigned them to the lenders' trust. Thus if the OTEC plant is producing a revenue stream, claimants may have to chase it. If the OTEC plant is defunct, both principal assets of the owners' trust and Energy Platform Corporation have disappeared. To what extent could claimants reach the general assets of the beneficial owners, the financial institutions putting up the equity? [They would have contractual recourse back to Ocean Energy Corporation and its parent corporation(s).] Or to what extent would the claimants' only hope be the Ocean Energy Corporation?

The outcome of these issues might depend on the degree to which maritime law was seen to be the appropriate body of principles. The financial institutions might argue that the governing concept is that admiralty is appropriate and that the lease is analogous to a demise charter. Under a demise charter, the charterer (i.e., Ocean Energy) is considered the owner *pro hac vice* and therefore responsible for the *in personam* liabilities arising from the structure's operation.[94] Whether the careful and clear placing of ownership in the owners' trustee for tax purposes, cemented by the IRS rulings (see Chapter 11), would defeat this argument is unclear.[95] In the *Torrey Canyon* case,[96] the traditional time charter format had been used, in part in order to establish that the third-party corporation was not a demise charter. Yet the appeals court remanded the case with instructions to the district court to let the claimants proceed against Union Oil Company, the operator, because the third-party corporation "owner" was a total shell. The case was settled before trial.

A closely related issue would be whether Ocean Energy Corporation, the owners' trustee, or the owners could move to limit their liability to the value of the structure, which would be zero if it had sunk, or for exoneration, under the Limitation of Liability Act (46 U.S.C. § 183, *et seq.*). Created in 1851 to provide a better competitive position for U.S. shipping, the statute provides that

> [T]he liability of the owner of any vessel . . . shall not . . . exceed the amount or value of the interest of such owner in such vessel[97]

The kind of analysis provided above as to the OTEC plant's qualification as a vessel would probably indicate that it would qualify. However, vessels which do not carry passengers have not been included under § 183(b)-(e), which applies to loss of life and injury liability, creating a fund and a limitation for liability arising for loss of lives. Whether OTEC would qualify naturally would be a new question. If it did, the standing of Ocean Energy Corporation as an owner would be a critical issue. Demise charters stand as owners, as indicated above, and so qualify for the limitation. But whether the analogy would be allowed to govern is, again, very arguable. In the *Torrey Canyon* case, Union Oil agreed that even if it was not a charterer of the nature accorded limitation rights under § 186, it was nevertheless an owner under § 183(a). The lower court agreed. Without deciding the point, the appeals court did pave the way for claimants to sue Union Oil on claims unrelated to navigation of the ship and without reference to the limitation statute. Thus it seemed to be opening up the prospect of unlimited liability for the operator-owner.[98]

Thoughtful policy consideration of these issues, once OTEC plants were proved, would seek to balance the desirability of attracting financial institutions as funding sources of OTEC, without undue exposure, and the social need to protect those who may suffer undue harm as a result of an OTEC mishap—or even routine operations. Lawyers and others are prone to designing means of financing and launching new structures while leaving issues of liability to be untangled after a catastrophe occurs. If the operator is reachable and is an entity of substance commensurate to the substance of the equipment it employs and/or is required to insure adequately, frequently the issues raised here would be equitably resolved. The operator—and other parties as well—ought not to be able to escape liability as a by-product of the financing process.

Another possibility for handling liability claims might be legislation modeled on the Price-Anderson Act,[99] providing government indemnity for claims over and above set levels of private insurance coverage. The possibility of extensive claims arising from a catastrophic nuclear incident, the deterrent that prospect raised to private participation in the U.S. atomic energy program, and the need for a clearly available forum for expeditious handling of claims were believed so serious that this country in 1957 passed legislation establishing a program of private insurance coupled with governmental indemnity. In OTEC's case, the act's blanket-coverage concept for indemnity seems considerably less applicable for OTEC than for nuclear energy. The implementation of two international conventions on oil pollution which provide for liability funds,[100] if adapted to OTEC problems, would seem an alternative means of handling likely environmental liability risk. And current discussion over the extension of Price-Anderson's life and readjustment of the private insurance and government indemnity levels illustrate the difficulties of building in this solution as a permanent concept. But the model is one which future planners might bear in mind.

In summary, should subsidization of OTEC become U.S. policy, it would appear better to accomplish that result by direct participation, loans, or guarantee of obligations than by limiting the owners' liability at the expense of those harmed or by providing blanket government indemnification.

There are other liability questions arising from the financing of OTEC plants which can be only mentioned here, but which call for further attention if OTEC reaches commercial viability. For example, if OTEC financing follows the preferred ship mortgage patterns (see Chapter 11), should the usual priority in any way be altered statutorily?

If the federal government becomes involved as a source of finance other than as a guarantor under a government-assisted ship mortgage bond (Title XI mortgage), will it be liable for any claims other than its immediate funding exposure?

Personal Injury and Death

The rapid expansion during the last several decades of personal injury cases in the United States, onshore as well as at sea, has been noted earlier. Remedies for wrongful death at sea have developed from a combination of general maritime law and statutory law. The field comprises, probably, the most frequently litigated area of maritime law.[101]

A seaman now normally has three remedies for personal injury. Two derive from ancient maritime law concepts. The first of these, maintenance and cure, is essentially a liability without fault of the shipowner to provide support for injuries or sickness which occur while the seaman is signed on. It is therefore based on the employment relationship. Since liabilities arising for maintenance and cure run against both the ship and the shipowner, a maritime lien is created and rights of recovery may be enforced *in rem* and *in personam.*

Second, a shipowner's responsibility to maintain a seaworthy vessel has created a liability for unseaworthiness under general maritime law. Today, this remedy is one existing irrespective of whether or not the shipowner or his agents are negligent, although originally this responsibility was much more narrowly construed. Under the liability explosion of the last few decades, both onshore and at sea, the doctrine of unseaworthiness has become one of absolute liability without fault. As in maintenance and cure, it runs against the shipowner *in personam* or against the ship *in rem.* Unlike maintenance and cure, it was not thought to be dependent upon the employment relationship. But it applies only to seamen or those who qualify as such.

Half a century earlier, the seaman's chances of recovery were more limited. Following the case of *The Osceola* in 1903,[102] seamen did not have the right to recover for injuries resulting from the negligence of the master or crew of a vessel. And the doctrine of unseaworthiness was confined to the "structure of the ship and the adequacy of her equipment and furnishings."[103] To cure the absence of a remedy for negligence of master and crew, Congress passed the Jones Act[104] as part of the Merchant Marine Act of 1920.[105] It provides a statutory remedy, running against the employer, for recovery for injury or death caused by negligence of those in charge of the vessel. The case law has developed so that this remedy can be brought only *in personam*, not *in rem* against the ship. As a result, recovery here is subordinate to all maritime lien claims which might be brought against the vessel. The statute incorporates the Federal Employers' Liability Act as the underlying legislation of the Jones Act. There have been several criticisms of the drafting of the act and resulting ambiguities. Over time these three remedies have come to be included together in nearly all personal injury suits of seamen, and a great deal of evolutionary confusion has taken place.

For example, the issue of who is a seaman to be covered by the act and the two common law remedies has required extensive judicial elaboration. By the late 1950s it came to be apparent that "any worker whose duties had the slightest context on navigable waters was entitled to a jury verdict on his seaman status and a jury finding that he was a seaman could not be set aside." Further, the cases "strongly suggested that it made not the slightest difference whether the 'structure' was in any sense a vessel in or even capable of navigation."[106] This trend was commented on in the section Applicability of Maritime Law to Nontraditional Uses of the Ocean.

The current practice of bringing all three counts together draws attention, where they exist, to the differences among the nature of the three counts. For example, some district court cases have said that maintenance and cure was not subject to limitation of liability. If maintenance and cure is not, should the Jones Act count or an unseaworthiness count be?[107]

As noted above, because earlier court cases hold that there is no *in rem* action on a Jones Act count, it would fall to the bottom of the list in any distribution. Query: Does its being joined with a maintenance and cure and seaworthiness count restore the recovery to a priority lien situation?[108] Further, Jones Act actions have been thought to run only to the employer. But we have seen that in a complicated operational structure such as today's, the operator of the structure may not be the employer or even the owner. A case has dismissed a Jones Act and maintenance and cure count against both an employer who is not the operator of the structure and the operator who is not the plaintiff's employer.[109] Query further: In such a situation, if neither Jones Act nor maintenance and cûre counts were applicable, could the ship nevertheless be gone against in an *in rem* action under maintenance and cure?

These issues are illustrative of those remaining open and ambiguous under current maritime law dealing with vessels. An additional layer of complexity is added when one considers that this present law may be applied to a wholly new structure, a new era of ocean use, represented by OTEC and other future structure-based uses of the ocean. Should even the concepts of *in rem* proceedings and maritime liens be applied across the board to these new uses?

Wrongful Death. Remedies for wrongful death on an OTEC structure would, under existing law, present an equally complicated picture. In 1886, *The Harrisburg*[110] set the policy that no remedy existed under general maritime law for wrongful death on the ocean. The recourse of the family of a victim was to state law—in this case the state in which the ship's owners were incorporated. Then or subsequently all states passed laws providing for recovery.[111] In 1920,

Congress enacted both the Death on the High Seas Act (DOHSA)[112] and the Jones Act, referred to above, which included recovery for wrongful death in addition to personal injury, DOHSA provided a remedy for death "caused by wrongful act, neglect or default occurring on the high seas beyond a marine league from the shore of any State [territory or dependency]."[113]

The decisional history dealing with DOHSA, the Jones Act, and the fifty state laws has resulted, to put it mildly, in a state of utter confusion. Differences as to whom the decedent might be (a seaman or anybody else), who the plaintiffs were (direct family or more removed), whether the claim was under a death count (i.e., rights of survivors) or a survival count (rights of the victim after the injury but before death), and place of the accident (high seas or territorial waters) all had an effect on the outcome.

In 1970, the Supreme Court issued a decision which may well reduce all the above considerations to background noise. In *Moragne v. States Marine Lines, Inc.,*[114] the Court held that it was creating a remedy for wrongful death under general, i.e., nonstatutory, maritime law. The case involved an accident in Florida's territorial waters, but its language reached beyond. Gilmore and Black summarize the case's impact:

> The Moragne remedy covers deaths within the territorial waters as well as deaths on the high seas. The remedy provides recovery for deaths caused by negligence as well as for deaths caused by unseaworthiness although in the latter case the decedent must have been a person (e.g. a Jones Act seaman) entitled to the warranty of unseaworthiness.[115]

If Moragne is in fact extended to the high seas and is developed to meet its full potential, there would be a remedy available for wrongful deaths on an OTEC structure in which the complications and inconsistencies evolved in the decisional interpretations of the Jones Act and DOHSA would be absent. If the case does not develop into its potential, Congress might seek to incorporate its initial principles into legislation dealing with OTEC and/or other offshore uses.

Liabilities Arising from Navigational Operations

An OTEC structure, in certain modes of operation, will be in navigational status. While an OTEC device is moving in the water, or even if it is stationary, it probably will be subject to the traditional sources of standards of correct navigational action, e.g., statutory rules of navigation, other statutes, local customs not contradicting the above, and requirements of good seamanship and due care. It is reasonably clear that maritime law would apply in collision cases. But several questions arise.

What difference, if any, would it make if its application was through the Maritime Torts Extension Act rather than as between the two vessels? Of what, if any, navigation responsibility would an OTEC structure be relieved if it were surrounded by a safety zone? How large should the safety zone be to insulate an OTEC plant effectively, say, from collision with a supertanker doing 20 kn? Should salvage principles apply to an OTEC structure if it suffers a casualty?

Liabilities of Designers, Constructors, and Service Contractors

Another category of issues would arise from OTEC's relationships with builders, designers, services, etc.

Should disputes arising from a construction contract of an OTEC plant be handled in the same legal framework as other contracted issues traditionally falling within maritime law? Does it make a significant difference?

Should a contract design characteristic which subsequently results in harm to persons, the environment, etc., form a basis for recovery?

Should liens apply to construction contracts if they apply to other aspects of OTEC plant liability?

Should classification or society approval of construction designs for OTEC structures be required? How should such approval and Coast Guard or other certification or administrative approval weigh in any subsequent liability challenge bearing on design adequacy?

Should liens against the OTEC structure be available to claimants in contractual claims arising from services, etc., provided as part of OTEC operations?

Environmental Law and Maintenance of Public Order

Two further categories of liability issues are only signaled here, but would require further analysis in any comprehensive legislation concerning liability.

Should federal environmental or state statutes or both be extended to cover OTEC operations within economic zone jurisdiction?

Should federal or state employer-employee relations laws pertaining to discipline be extended, or should traditional plus statutory (e.g., 46 U.S.C. § 704) maritime law be applied? Should new statutory provisions for responsibility on board, similar to those enacted in England, be considered?

Should there be a civil cause of action or criminal offenses for intentional acts harmful to an OTEC structure?

Approaches to the Problem of Legislation regarding OTEC Liability

The U.S. Congress at some point will undoubtedly enact new legislation dealing with OTEC, either specifically or as one of a class of offshore uses. That legislation ought to include consideration of the issues treated in this chapter. Two questions will be pertinent, over and above those of forum discussed in the section A Forum for OTEC Liability and those substantive issues raised illustratively in the section Liability Issues. The first is the body or bodies of law Congress makes applicable. The second is the degree of policy direction which is detailed in the legislation. The approaches to these questions taken in two prior pieces of legislation—the Outer Continental Shelf Lands Act (OCSLA), with over 20 years' accumulated experience, and the Deepwater Port Act (DPA) of 1974—provide some insights. Additionally, the manner proposed for handling similar issues in an international treaty covering Ocean Data Acquisition Systems (ODAS) suggests possible routes such legislation might take.

The Outer Continental Shelf Lands Act and the Deepwater Port Act of 1974

In laying the framework for the first substantial new offshore technology over 20 years ago, Congress provided in OCSLA that:

> The Constitution and laws and civil and political jurisdiction of the United States are extended to the subsoil and seabed of the outer Continental Shelf and to all artificial islands and fixed structures which may be erected thereon for the purposes of exploring for [and exploitation of] resources therefrom, to the same extent as if the outer Continental Shelf were an area of exclusive federal jurisdiction located within a state.[116]

and further that:

> To the extent that they are applicable and not inconsistent with this subchapter or with other federal laws . . . the civil and criminal laws of each adjacent State . . . are declared to be the law of the United States for that portion . . . of the outer Continental Shelf . . . which would be within the area of the State if its boundaries were extended seaward. . . . All of such applicable laws shall be administered and enforced by the appropriate officers and courts of the United States.[117]

The act also provided the federal district courts with original jurisdiction over cases and controversies "arising out of or in connection with any operations conducted on the outer Continental Shelf"[118] and made applicable the Longshoremen's and Harbor Workers' Compensation Act.[119]

Twenty years after the passage of the continental shelf acts, which responded to the technology of platform drilling and production, the United States was again faced with the need to respond legislatively to new ocean technology—the supertanker and its deepwater port. The analogies to the OTEC situation are worth noting: location beyond present territorial seas; physical attachment to the seabed, but no exploitation connection; vessel traffic tying up and departing; polluting risks; and the uncertain status of a floating platform.

The DPA[120] follows the main design of OCSLA in extending to deepwater ports (which by definition are beyond territorial waters)[121] and to "activities connected, associated, or potentially interfering with the use or operation of any such port" the U.S. Constitution, laws, and treaties "in the same manner as if such port were an area of exclusive federal jurisdiction located within a State."[122] It also made the law of the nearest adjacent coastal state law applicable:

> [T]he law of the nearest adjacent coastal State . . . the law of the United States, and shall apply to any deepwater port licensed pursuant to this Act, to the extent applicable and not inconsistent with any provision or regulation under the Act or other federal laws and regulations . . . All such applicable laws shall be administered and enforced by the appropriate officers and courts of the United States.[123]

Further, U.S. district courts shall have original jurisdiction of cases and controversies arising out of or in connection with the construction and operation of deepwater ports.[124] These features are essentially common with OCSLA. However, beyond these points lie a multitude of differences, as examined below.

Choices among Bodies of U.S. Law

The coupling in both of the above acts of federal civil and criminal laws and the state's own civil and criminal laws, where applicable and not inconsistent with federal law, raises the question as to why state law applicability was necessary or advantageous. As pointed out in the very first section, the residual civil and criminal law of this country is in the state domain. In the historical sense, there is no federal common law, although a body of tort and contract law has been built up in the federal courts. If offshore continental shelf operations, deepwater ports, and OTEC structures or other offshore space were to have U.S. civil and criminal laws applied to them as if they were a federal enclave within a state, and no adjacent state law were made applicable, there would be innumerable gaps. Whatever law has been built up at the federal level dealing with tort liability is not a complete system. In order to close the gaps and provide a complete civil and criminal regime, Congress made adjacent state law assimilated in the offshore space.

What does the application and assimilation of state law do to the application of maritime law? The OCSLA applied federal laws without specific reference to admiralty. It said nothing about the status of "artificial islands and fixed structures" for purposes of maritime law. One would assume that maritime law is included in the phrase, *U.S. civil and criminal laws.* Further, the statutory language referring to federal law as if it were an area of exclusive federal jurisdiction *within a state* presumably does not make maritime law any less applicable. The substance and jurisdictional reach arise as a function of both the federal judicial jurisdictional grant in 28 U.S.C. § 1333 and the maritime nature of the subject matter (see the sections Introduction, and Applicability of Maritime Law to Nontraditional Uses of the Ocean).

Also the statutory language indicated that when inconsistent with it, state law is subordinate to federal. This supposition is reinforced by the concept of federal uniformity in maritime matters.[125] Thus it might be assumed that maritime law would be the primary residual law on the continental shelf.

However, the act's legislative history spoke to the issue, as was seen in the section Applicability of Maritime Law to Nontraditional Uses of the Ocean. Congress considered and rejected using maritime law as the basic body of law, fearing that it was not in itself complete.[126] This, indeed, appears to be a principal reason for the assimilation of adjacent state law. Any automatic assumptions about the relationships of maritime and state law may have been further weakened by a 1973 Supreme Court case. In *Askew v. The American Waterways Operators, Inc.*,[127] the Court upheld a Florida law establishing liability for oil spills in territorial waters in potential conflict with the Federal Water Pollution Control Act.

The questions raised in the above paragraphs make it all the more desirable that in the future Congress be as explicit as possible as to the division of labor between both statutory and general maritime law on the one hand and the adjacent state law to be assimilated on the other.

In contrast to the OSCLA, the DPA, without mentioning admiralty, in effect blends some of its features in with a host of other details which give shape to the liability aspects of the regulatory regime. The difference in approach raises questions as to the best approach in future legislation as to the amount of detail optimally included by the Congress.

Comparison of the Two Acts as to Degree of Policy Direction in the Legislation

Congress's intent to favor adjacent state law over maritime law in the case of artificial islands and fixed structures became significant only 17 years later in the Rodrigue case.[128] Given the ferment during the interim among the federal courts concerning liability for injury and death[129] and the unanticipated advances in offshore technology, it is not surprising that the legal status of offshore structures is chaotic today.

The contrast in detail between OCSLA and the DPA of 1974 is marked, and must represent a determined attempt by the Executive and Congress to clarify many of the ambiguities or make policy decisions on disputed points in the public and private remedy domain that have plagued the maritime area. The provisions are worth noting, for if the courts uphold them and breathe life into them and if they grow healthily, they may provide guides to future OTEC legislation.

The regulatory design is a licensing one, with very broad regulatory power vested in the Secretary of Transportation[130] and with the threat of revocation of licenses[131] and of civil suits[132] to enforce compliance. The current concern for the ocean environment is reflected in the act, with the discharge of any oil being forbidden.[133] In addition, compliance with or reference to the Clean Air Act,[134] Federal Water Pollution Control Act,[135] Marine Protection, Research and Sanctuaries Act,[136] National Environmental Policy Act,[137] Interstate Commerce Act,[138] and Coastal Zone Management Act[139] is built in.

The sections dealing with remedies (§ 15), citizen civil action (§ 16), and liability (§ 18), however, are the major concerns here, for they address many issues relating to the status of maritime law in OTEC.

Section 15 provides fine and/or imprisonment for violation of the act or any rule, order, or regulation under it.[140] But it also provides the government itself with an equity-type relief: the Secretary of Transportation can issue an order requiring compliance and subsequently get the Attorney General to seek appropriate relief in a civil action, including an injunction or civil penalty, to ensure compliance.[141] And the Secretary of Transportation can request the Attorney General to seek equitable relief to redress any violation of the act or regulation directly, without the compliance order.[142] In addition, any person may commence a civil action for equitable relief on his own behalf against another in violation of the act or any condition without regard to amount in controversy or diversity.[143] In these two provisions the act provides for equity relief, which is missing normally in federal admiralty law.

The act provides for *in rem* liability of vessels used in a violation, thus incorporating a critical feature of admiralty.[144]

In terms of the state-federal jurisdictional question, the citizen civil action section also provides a modern-day equivalent to the saving-to-suitor clause, implicitly allowing state law remedies.[145] And the liability section states that it "shall not be interpreted to preempt the field of liability or to preclude any State from imposing additional requirements or liability for any discharge of oil. . . ."[146] Thus a significant potential issue is at least addressed, if not laid to rest.

By including in the district courts original jurisdiction over cases and controversies over the construction as well as the operations of deepwater ports,[147] the act, without explicitly saying so, may have pushed construction contracts closer to admiralty, or at least brought that class of cases together under the same roof with subjects and procedures which would now be considered maritime in nature. The push is further reinforced by the extensive design and other construction data to be submitted in the license application.[148]

In a lengthy and detailed section on liability for discharge of oil (§ 18), the act specifically deals with points which are characterized by either their presence or absence in maritime law. Thus it provides for joint and several liability and liability without regard to fault [§ 18(h)(1) and (2)], saving of rights against third parties [§ 18(h)(4)], and statute of limitations [§ 18(j)(2) and (3)].

Draft Articles Prepared by the Group of Experts on the Private Law Aspects of Ocean Data Acquisition Systems, Aids, and Devices, (ODAS)

In May 1974, a group of experts gathered by UNESCO's Intergovernmental Oceanographic Commission (IOC) prepared draft articles pertaining to the private law or liability aspects of ocean data acquisition systems[149]—those structures or installations used at sea to collect, store, and transmit marine environment or atmospheric data. The draft articles were designed as a component for a proposed convention on ODAS drawn up under the authority of the IOC in the 1960s.[150]

These articles offer still another legislative approach in a closely analogous field to the liability issues reviewed in this chapter. Conceptually, they might be defined as standing somewhere between the detailed, wide-ranging approach of the DPA and the minimal reference to applicable law found in the OCSLA. The approach, summarized below, is to state the basic principles to be applied to the most critical issues of responsibility and liability among the parties.

Thus proportional or relative fault is declared the basis for assessing liability for loss or damage in cases of collision, unless no determination of fault is made because the loss or damage is due to accident, *force majeure*, or is in doubt. If so, the draft provides that losses will be borne by those who suffer them.[151] Similar provisions hold in torts other than collision, except that it is provided that national legislation can override the convention. In two cases, the use of radioactive materials in an ODAS and failure to display a signal as provided in the parent convention, the owner or operator shall be held liable without regard to fault.[152]

Contractual rights and obligations—relating to maintenance, repair, towage, contract salvage, or modification of an ODAS—will be governed by the law of the place where the work is performed or, failing its determination, by the law of the forum.[153] Parties or "applicable legislation" can provide otherwise. The law of the state in which the ODAS is registered shall govern social legislation such as conditions of employment, compensation, and the like.[154]

These examples illustrate the statement of guiding principles found in the articles. Similar guides are set out for limitation of liability (a fixed monetary amount),[155] liens and attachments (they do not apply if the ODAS has an appointed representative for service of process and court appearance instead), provision for joint and several liability in cases of death or personal injuries,[156] exoneration for the owner when a person is damaged by reason of his own act or omission done with intent to cause damage,[157] compensation for putting one's vessel or other property in jeopardy to avoid harm to an ODAS,[158] civil liability for interfering with an ODAS and consequently harming it,[159] and salvage.[160]

By stating these principles, a regime is outlined to handle many of the basic issues of maritime law, or at least give direction as to what law will apply.

Conclusion

This chapter has reviewed only a portion of the liability issues likely to arise if OTEC is developed and becomes a significant source of offshore energy by the end of the century. Yet they are typical in that they signal the kinds of problems which can be anticipated and the kinds of decision that policymakers will have to make.

The liability issues will be resolved if cases are brought to a forum. It is most likely that there will be a forum and a body of law for bringing those cases. Past experience suggests that federal district courts will have original jurisdiction over them. Whether the predominant body of law ought to be maritime law, adjacent state law, or a new federally created set of principles remains unclear and requires extensive analysis. However, the option of maritime laws being applicable to OTEC and similar structures exists under current legal interpretation.

The areas of financing and ownership patterns and of personal injury and death claims illustrate two important points. First, laws and legal usages designed to get something done (in this case get the financing of OTEC accomplished through institutions created or adapted to meet financial and economic reality) tend to come into being while issues concerning liability (in this case what happens if there is a failure) tend to lag behind. Today lawyers who put together the legal institutions to accomplish the financing of large offshore structures analogous to OTEC really are not sure of what the resolution of liability issues would be, should one be lost. From the plaintiff lawyer's viewpoint, there are fundamental policy issues which frequently play second fiddle.

Second, existing law (in this case particularly maritime law), although having the advantages of being available and providing a cohesive body of law, contains within it many inconsistencies, outmoded principles established for another era, and ambiguities. These deficiencies are compounded when the interplay between state and general or statutory maritime law is considered. Probably no single body of existing law is ideal or should be applied in blanket fashion.

Rather, the existing principles should be examined; trouble spots or inconsistencies should be questioned; and where advantageous, the advent of a new era in offshore technology should be taken advantage of to improve the body of law.

In order to accomplish this, Congress would have to go into fair detail as to policies and principles when it begins to write new offshore law. The OCSLA, the first applicable model, did not do this. The DPA attempted to fill in more of the interstices rather than relying only on the application of adjacent state law. The statement of general principles such as in the proposed ODAS convention would seem to suggest useful components in any new legislation.

Finally, the timing of such legislation must be realistic in terms of the development of the technology itself. There are strong arguments for promotional or financing laws to precede regulatory or residual law legislation. But as complex as these latter problems are, they ought not to be left until the eleventh hour.

Notes

1. *See* Chapter 1 for technical description and analysis.

2. Lockheed and TRW have each completed very preliminary prototype design configurations for an OTEC plant based on state-of-the-art technology. For purposes of this chapter, these two models will be used interchangeably unless a characteristic of one requires specification.

3. 43 U.S.C. § 1331 *et seq.*

4. *See also* Gilmore and Black, *The Law of Admiralty* (Mineola, N.Y.: Foundation Press, 1975) 2nd ed. 986 [hereinafter cited as Gilmore and Black].

5. 28 U.S.C.A. § 1333.

6. One pertinent model, the jurisdictional grants in the Outer Continental Shelf Lands Act, 33 U.S.C. § 1133, is discussed below.

7. Gilmore and Black 1-2. The following characterization of maritime law is drawn from the first chapter of Gilmore and Black.

8. Id. at 11.

9. Id. at 47.

10. Id. at 19.

11. For a preliminary model of relationships between law and technology, *see* Nyhart, "The Interplay Between Technology and Law in Deep Seabed Mining Issues" (forthcoming), *U. Va. J. of Int'l L.*

12. U.S. Const., Art. I, § 2.

13. 1 Stat. 76-77.

14. The concept is now embodied in 28 U.S.C.A. § 1333 in somewhat altered language: "saving to suitors in all cases all other remedies to which they are otherwise entitled."

15. Gilmore and Black 19.

16. DeLovio v. Boit, 7 F.Cas. 418 (No. 3776), at 444 (1815).

17. For examples of claims considered maritime torts, *see* Gilmore and Black 23.

18. Id. at 32-33.

19. Id.

20. Id. at 283.

21. 46 U.S.C. §§ 971-75.

22. 46 U.S.C. §§ 911 *et seq.*

23. 46 U.S.C. § 11 *et seq.*

24. 46 U.S.C. § 181 *et seq.*

25. Executive Jet Aviation, Inc. v. City of Cleveland, Ohio, 409 U.S. 249 (1972).

26. In essence, the Court was modifying an 1866 statement that "Every species of tort, however occurring, and whether on board a vessel or not, if upon the high seas or navigable waters, is of admiralty cognizance." The Plymouth, 70 U.S. (3 Wall.) 20, 36 (1866).

27. 1 U.S.C. § 3; this definition was originally enacted as part of the Act of July 18, 1866, Ch. 201 § 1, 14 Stat. 178, dealing with prevention of smuggling, and incorporated into the Revised Statutes as § 3 of the Act of July 30, 1947, e.g., Title 1 of the U.S.C. 61 Stat. 663; 1 U.S.C. § 3; 3 *ALR Fed.* 890. It has served as a model in the *Code of Federal Regulations* dealing with documentation and measurement of vessels, 46 C.F.R. 66.03-5 and in the National Boating Safety Act, 46 U.S.C. 145, § 1452(2). Both of the latter definitions excluded aircraft. Gilmore and Black reflect the transportation theme: "In the dealings of the courts, a single clear test is hard to discern; perhaps the best approximation would be to say that the term "vessel" is applied to floating structures capable of transporting something over the water" (at 33).

28. As a category of structure, barges have almost uniformly been held to fall within maritime law. Cases cited, 44 *Words and Phrases* 240. For statutory basis of barge, *see* Seagoing Barge Act, 46 U.S.C. 395.

29. Similarly, dredges as a category appear to have generally been classified as vessels. Cases cited, 44 *Words and Phrases* 241-3; *contra* Bartlett v. Steam Dredge No. 14, 64 N.W. 951, 952 (for purposes of maritime lien), U.S. v. Dunbar, 67 F.783, 784 (for customs purposes).

30. The Sallie, 167 F.880 (seaman's lien for wages); Patton-Tully Transp. Co. v. Turner, C.C.A. Tenn., 269 F.334, 336 (limitation of liability).

31. Lawrence v. Flatboat, 84 F.200, 201; George Leary Const. Co. v. Matson, C.C.A. Va., 272 F.461, 462 (personal injury); *contra* Pile Driver E.O.A., 69 F.1005, 1007.

32. *Infra*, section Lower Court Application of Maritime Law.

33. The Hezekiah Baldwin, 12 Fed. Cas. 93.

34. The O'Boyle No. 1, D.C.N.Y., 64 F. Supp. 378, 382 (limitation of liability).

35. Charles Barnes Co. v. One Dredge Boat, 169 F.895, 896.

36. The Scorpio, 181 F.2d. 356 (1950).

37. *See* 44 *Words and Phrases* 247, citing cases covering general maritime jurisdiction, Longshoremen's Compensation Act, and limitation of liability.

38. M/V Marifax v. McCrory, 381 F.2d 909 (C.A.S. Fla. 1968).

39. For purposes of maritime liens for supplies and repairs under U.S.C. 971, *see* 3 *ALR Fed.* 895.

40. *See* text accompanying note 94 *infra.*

41. *See* text accompanying note 65 *infra.*

42. *See* text accompanying note 72 *infra.*

43. The Pirate Ship (1927, D.C. La) 21 F.2d 231.

44. The Showboat, 47 F.2d 382 (1939 D.C. Mass.) (maritime lien for repairs and supplies).

45. The Ark, 17 F.2d 446 (1926, D.C. Fla).

46. Pleason v. Gulfport Shipbuilding Corp., 221 F.2d 621 (1955, C.A.S. Tex).

47. *See* 3 *ALR Fed.* 882.

48. Id.

49. Id.

50. *See* cases cited, 44 *Words and Phrases* 240.

51. 3 *ALR Fed.* 882.

52. Id.

53. Saylor v. Taylor, 77 F.476 (1896).

54. Cope v. Vallette Dry-Dock Company, 119 U.S. 625 (1887).

55. Rodrigue *et al.* v. Aetna Casualty and Surety Co. *et al.*, 395 U.S. 352.

56. Kenny v. City of New York, 108 F.2d 958 (1940).

57. Cook v. Belden Concrete Products, Inc., 472 F.2d 999 (1973).

58. *See* reference cited *supra* note 47.

59. Id.

60. Id.

61. Id.

62. *Supra* note 55.

63. Offshore Company v. Robison, 266 F.2d 769.

64. Sims v. Marine Catering Service, Inc., 217 F.Supp. 511 (1963).

65. Marine Drilling Co. v. Austin, 363 F.2d 579 (5th Cir. 1966).

66. Producers Drilling Co. v. Gray, 361 F.2d 432 (5th Cir. 1966).

67. Hicks v. Ocean Drilling and Exploration Co., 512 F.2d 817 (1975).

68. Id. at 821:

The primary meaning of the term "vessel," is any water craft or other contrivance used or capable of being used as a means of transportation on water. The term "vessel" may also include, however, various special purpose crafts, such as barges or dredges, which do not operate as a vessel for transportation, but rather, serve as moveable floating bases that may be submerged, so as to rest on the bottom and be used for stationary operations, such as drilling or dredging. Now, although mere flotation may not in and of itself be sufficient to make a structure a vessel, if a structure is buoyant and capable of being floated from one location to another, it may be found to be a vessel, even though it may have remained in one place for a long time and even though there are no plans to move it in the foreseeable future. Now, in considering whether a special purpose craft is a vessel, the determinative factors are the purposes for which the craft was constructed and the business in which it is engaged. That is, was the craft designed for and used in navigation or commerce? A craft not designed for navigation and commerce, however, may be classified as a vessel, if at the time of the accident, it had actually been engaged in navigation or commerce. Now, in considering whether a special purpose craft is a vessel, the manner in which a party or parties may have referred to or denominated such craft in contracts or other documents is not necessarily determinative of its status as a vessel, but is simply a factor for you to consider along with all the other evidence.

69. The specific inclusion of mobile and built-up platforms may be an anomaly in light of the actual practices referred to. If so, it appears compounded by the definition of a mobile platform: "an artificial island or fixed structure, which itself includes as an integral part of itself features which permit it to be moved as an entity from position to position and to be fixed to or submerged onto the seabed" (33 C.F.R. 140.10-30).

70. For financing purposes, platforms also have been treated analogously to vessels. Perhaps only an earthen artificial island built of dredged material is the only permanently affixed structure, since the financing terms of a leverage leased gravity based platform, floated to the site, then sunk to rest on the bottom, require it to be capable of being refloated for reassignment at the expiration of its 25-year lease.

71. Rodrigue, *supra* note 55.

72. Id.

73. Chevron Oil Co. v. Huson, 404 U.S. 97 (1971).

74. Louviere v. Shell Oil Co., 515 F.2d 571 (1975). In Dickerson v. Continental Oil Co., 449 F.2d 1209 (1971), the Court described the structure as "an offshore stationary (fixed) drilling platform," thereby emphasizing its nature.

75. Day v. Ocean Drilling and Exploration Co., 353 F.Supp. 1350 (1973).

76. Hicks, *supra* note 67.

77. Nations v. Morris, 483 F.2d 557 (1973).

78. Continental Oil v. London Steam, 417 F.2d 1030 (1969).

79. Oppen v. Aetna Insurance Co., 485 F.2d 252 (1973).

80. The view that expansion of maritime tort laws should be slowed was also recognized in recent amendments by Congress to the Longshoremen and Harbor Workers Act, altering two of the Supreme Court's positions on the rights of harbor workers to recover from shipowners. *See* Gilmore and Black 423.

81. Davis v. City of Jacksonville Beach, 251 F.Supp. 327 (1965).

82. Madole v. Johnson, 241 Supp. 279 (1966).

83. Executive Jet, *supra* note 26.

84. Id.

85. *Supra* note 53.

86. It is assumed that the OTEC is a U.S. owned vessel, and, if documented, is a U.S. flag one.

87. "The district courts shall have original jurisdiction, exclusive of the courts of the states of:

> (1)Any civil case of admiralty or maritime jurisdiction, saving to suitors in all cases all other remedies to which they are otherwise entitled" (28 U.S.C. 1333).

88. Id.

89. Gilmore and Black 37.

90. 33 U.S.C. § 1331 *et seq.*

91. 33 U.S.C. § 1501.

92. *See,* e.g., the Settlement of Disputes section of the Informal Single Negotiating Text, Third U.S. Law of the Sea Conference, SD. gp/2d Session/No. 1/Rev. 5, 1 May 1975.

93. The model followed here provides an additional entity, Energy Platform Corp., in order to provide an issuer of equipment notes acceptable under New York security laws.

94. Gilmore and Black 242.

95. Id. at 841.

96. *In rem* the Complaint of Barracuda Tanker Corporation as Owner of the S/T Torrey Canyon and Union Oil Company of California for Exoneration from or Limitation of Liability. 409 F.2d 1013 (1969).

97. 46 U.S.C. § 183.

98. Gilmore and Black 841.

99. 42 U.S.C. § 2210, *et seq.*

100. International Convention on Civil Liability for Oil Pollution Damage (1969); and International Convention on the Establishment of an International Fund for Compensation for Oil Pollution Damage (1972).

101. In the discussion that follows, it is assumed, following the analysis of Section II, that OTEC would be considered a vessel for these purposes. *Generally, see* Gilmore and Black, Chapter VI, from which the following descriptions and characterizations were largely drawn. The author holds himself wholly responsible for any misinterpretations in this complex field.

102. The Osceola, 189 U.S. 158 (1903).

103. Gilmore and Black 277.

104. 46 U.S.C. § 688, *et seq.*

105. 41 Stat. 1007 (1920).

106. Gilmore and Black 333.

107. Id. at 341.

108. Id. at 342.

109. Mahramas v. American Export Isbrandtsen Lines, Inc., 475 F.2d 165.

110. 119 U.S. 199 (1886).

111. Gilmore and Black 359.

112. 46 U.S.C. §§ 761-768.

113. 46 U.S.C. § 761.

114. 398 U.S. 375 (1970).

115. Gilmore and Black 368.

116. *Supra* note 3, § 1333(a)(1).

117. Id. at § 1333(a)(2).

118. Id. at § 1333(b).

119. Id. at § 1333(c).

120. 33 U.S.C. § 1501.

121. Id. at § 3(10).

122. Id. at § 19(a)(1).

123. Id. at § 19(b).

124. Id. at § 19(e).

125. Southern Pacific Company v. Jensen, 244 U.S. 205 (1917).

126. *See* text accompanying note 55, *supra.*

127. Askew v. American Waterways Operators, Inc., 335 F.Supp. 1241, rehearing denied 411 U.S. 325 (1973).

128. *Supra* note 55.

129. The area of liability for personal injury and wrongful death is especially revealing as to future problems for OTEC, first in that the period of offshore oil development, has paralleled a time of voluminous litigation. There has been, therefore, a considerable opportunity for judicial expression on the subject of current offshore technology's place in maritime law. It was a period in

which the Supreme Court was judicially rewriting the law of maritime injury, to which there may now exist both a judicial and legislative backlash. It is an area which says something about reliance on judicial elaboration of congressional enactment in new ocean technology. *See* Gilmore and Black VI, *in passim.*

130. *Supra* note 120 at § 10.

131. Id. § 12.

132. Id. § 16.

133. Id. § 18(a)(1).

134. 42 U.S.C. § 1857.

135. 33 U.S.C. § 1251.

136. 33 U.S.C. § 1401.

137. 42 U.S.C. § 4321.

138. 49 U.S.C. § 1-27.

139. 16 U.S.C. § 1451.

140. *Supra* note 120 at § 15(a).

141. Id. § 15(b)(1), (3).

142. Id. § 15(c).

143. Id.

144. Id. § 15(d).

145. Id. § 16(e).

146. Id. § 18(k)(1).

147. *See* text accompanying note 124 *supra.*

148. *Supra* note 120 at § 5(c)(2).

149. Draft Articles Prepared by the Group of Experts on the Private Law Aspects of Ocean Data Acquisition Systems, Aids and Devices (ODAS), *Annex II*; First Session of the Group of Experts on Private Law Aspects of Ocean Data Acquisition Systems, Aids and Devices (ODAS), U.N. Doc. 10C/ODAS-LEG/3 (SC–74)/CONF. 676/COL.2), 1974 [hereinafter Draft Articles].

150. Annex III, Preliminary Draft Convention on Ocean Data Acquisition Systems, Aids and Devices (revised), U.N. Doc. SC–72/CONF. 85/8, Annex III.

151. Draft Articles § II.

152. Id. § I.

153. Id. § III.

154. Id. § IV.

155. Id. § XI.

156. Id. § VIIa.

157. Id. § VIIb.

158. Id. § VIII.

159. Id. § IX.

160. Id. § XII.

Ocean Thermal Energy Conversion Plants: Federal and State Regulatory Aspects

James C. Higgins, Jr.

Introduction

A fundamental legal question respecting the construction and operation of OTEC plants concerns the nature of the framework of applicable regulatory law. Should there be federal licensing of the plants? Should the federal government exercise exclusive authority, or should the states participate jointly? Should the states regulate the facilities to the exclusion of the federal government? Should either sovereign have a predominant role in licensing?

The discussion in this chapter applies mainly to the regulation of OTEC plants in the commercial development stage. Some of what is said would apply also to any prototype and demonstration phases which the Energy Research and Development Administration (ERDA) ordains as necessary for the proper development of this energy system, since in these phases some technical review and certification or licensing will be required.

Federal Jurisdiction

Need for Federal Control

OTEC plant operating conditions will require plant siting in tropical and subtropical waters adjacent to the United States. Because of siting constraints (e.g., ship traffic, water temperature, weather and climatic conditions, tides, currents, environmental prohibitions, bottom topography, seismic criteria, and alternate uses), the number of acceptable locations for the stationing of OTEC facilities will doubtless be quite limited. It is not inconceivable that acceptable sites will be found only on the continental shelf adjacent to a single state. If the technology for economical transmission of electric power is developed, however, the demand for power from the facilities will more than likely span several coastal states. It is therefore important that control of site allocation not be left to one state. The planning and allocation of sites for OTEC plants should be a matter requiring federal supervision.

There are additional reasons for federal control. Sited (as they will be) in or near shipping lanes (posing obstructions to navigation) well outside the 3-mi territorial sea and state jurisdiction, in national waters superjacent to the outer continental shelf (therefore making a new use of the continental shelf and water column in which the federal government has an exclusive interest), OTEC facilities are clearly of such national scope as to be subject to federal licensing. Nevertheless, as discussed in detail later, the states must be given a significant role in the planning and regulation of OTEC plants.

Precedents

Precedents exist for the regulation by the federal government of OTEC plants. Energy or energy-related facilities sited beyond the U.S. territorial sea have been made subject to federal control. Thus, Congress has provided for Department of Transportation (Coast Guard) regulation of deepwater ports.[1] Nuclear power plants constructed and operated under provisions of the Atomic Energy Act of 1954, as amended, require licensing by the Nuclear Regulatory Commission (NRC).[2] Hydroelectric plants on navigable waterways are constructed under the authority of Federal Power Commission licenses with Corps of Engineers concurrence.[3] Electric power facilities at the U.S. border for the transmission of power between the United States and a foreign country are licensed by the Federal Power Commission.[4]

Prompt production and delivery of clean, safe, and reliable domestic energy, as well as the development of energy sources, has become critically important for the United States in the 1970s, and doubtless will continue to be so for several decades. Assuming that a regulatory role for the federal government in the ocean thermal energy field is deemed necessary after completion of prototype and demonstration phases, it is important that Congress devise a licensing system that precludes unnecessary delay and duplication of administrative review. One-stop licensing in a single agency, streamlining the agency review, hearing, and appeal process, and combining (where not precluded by the need for a final design in the early commercial review stages) the licensing of construction and operation of the plants should be provided for. Most of these ideas were enacted by Congress for deepwater port licensing. OTEC plants will be prevented from making a worthwhile contribution to America's coastal energy requirements if unnecessary delays in obtaining government approvals during the commercial development phase are not precluded by similar legislative action.

Lead Agency

It is beyond the scope of this chapter to evaluate the various alternative licensing regimes available to Congress for application to OTEC facilities. A single, controlling agency should certainly be designated, however, as was done in the Deepwater Port Act (DPA) of 1974.[5] From the license applicant's viewpoint, a lead agency eliminates the burden of dealing individually with a host of administrative bodies. From the government's standpoint it centralizes control, scheduling, and review in one unit and provides a collecting point for the advice and consent of other departments vested with jurisdiction over ocean development. Under existing law, several candidates exist for the lead agency role. The Corps of Engineers, the Coast Guard, and the Interior Department could qualify. The National Oceanic and Atmospheric Administration and the Federal Energy Administration, each with its supporters in Congress, also could be vested with the responsibility. It will be up to Congress, however, to make this decision.[6]

Licensing Procedure. Under a one-stop licensing regime, as enacted for deep-water ports, one application would be submitted to the federal government by an applicant to construct and operate an OTEC plant. This application—to a lead agency—would "constitute an application for all Federal authorizations required for ownership, construction and operation of"[7] an OTEC plant.

Notice of receipt of an application is circulated to members of the public as well as federal and state governmental agencies by the licensing agency. A draft environmental impact statement is published, as required under Section 102(2)(c) of the National Environmental Policy Act,[8] and circulated to interested members of the public and federal and state governmental agencies for comment. The agency must consider the views of all persons commenting, and afford an opportunity for a public hearing. Each federal agency with responsibility in marine affairs, the environment, energy, safety, and so on must comment and, where required by law, issue (or deny) a license or certify that the construction and operation of the facility will (or will not) comply with appropriate provisions of federal law (these other agencies are discussed in the section of this chapter entitled Specific Federal Regulation). If the state's coastal zone management program has been approved by the Secretary of Commerce under the Coastal Zone Management Act of 1972, the applicant must furnish a certification that the proposed activity will be consistent with the program.[9] Among the matters considered by the agency in evaluating the application are the effect of the proposed project on navigation; fish and waterfowl; water quality; the general environment; beach erosion; historic, scenic, and recreational values; the economy; and the needs and welfare of the people.[10] The application must include a certificate of water quality under Section 401 of the Federal Water Pollution Control Act (FWPCA) from the state in which any plant-to-shore transmission lines are sited certifying that any dredging required to emplace the lines in the seabed will be in compliance with applicable effluent limitations and water quality standards.

A hearing is held allowing members of the public to appear and give evidence. Upon completion of the public hearing, receipt of all comments, and full evaluation by the agency staff, the agency publishes a final environmental impact statement and a decision on whether to issue the necessary permit. One permit would be issued for all proposed construction activities and, most likely, a second permit for operation. With one-stop licensing, the complete process of licensing an OTEC station might take as long as 12 to 18 months. A license would be valid for 20 to 40 years, and annual lease payments for the use of the seabed, water column, and airspace could be required.

Site Allocation. Because of siting constraints and a resulting shortage of acceptable sites, more than one company or authority may submit an application for the same plant location. Congress can anticipate such an eventuality, as it did with deepwater ports,[11] and provide for issuance of licenses by the administrative agency according to an order of priorities. An electric power company or authority in an immediately adjacent coastal state might be deemed to have first preference for the site, and an electric power company or authority serving customers in a less immediate or adjacent state might be second, for example. The agency may depart from this system of priorities if it determines that one of the proposed facilities would better serve the national interest. In making such a determination, the need for power in other states, environmental impact, as well as cost and delivery time could be used as criteria.[12]

Federal Jurisdiction on the Continental Shelf

Siting locations are important to the question of federal jurisdiction. Current schemes provide for siting of OTEC plants beyond the territorial sea. Some debate has traditionally centered upon the question of whether Congress extended the reach of U.S. jurisdiction to the lands of the continental shelf lying beyond the territorial sea in connection with the siting and operation of artificial islands and structures not erected for exploration or extraction of mineral resources. One relevant statute, Outer Continental Shelf Lands Act (OCSLA),[13] appears to extend such jurisdiction only for the purposes of mineral exploration and removal. However, the Corps of Engineers has asserted jurisdiction and permitting authority[14] over the siting of artificial islands and structures without regard to their use in exploration or extraction of resources, citing as authority Section 4(f) of the Outer Continental Shelf Lands Act.[15]

There are no cases clearly defining the extent of U.S. jurisdiction over nonextractive uses of the outer continental shelf.[16] Since OTEC plants would be nonextractive uses, sited beyond the territorial sea, there is less than clear legislative authority for federal establishment and total regulation of them. Accordingly, in addition to the reasons noted above regarding desirable administrative changes, it would seem prudent for Congress to enact legislation authorizing the construction, siting, and operation of these energy facilities, as it did in the DPA of 1974.[17]

Specific Federal Regulation

In the following sections of this chapter an analysis of specific federal regulatory requirements applicable to OTEC plants will be undertaken.

Army Corps of Engineers

Substantive Authority. Since siting of OTEC plants and laying of transmission cables (where an OTEC facility is used to generate electric power to shore) will occur in the ocean off the coast of the United States, the jurisdiction of the Corps of Engineers, Department of the Army, will be invoked. Under Section 10 of the River and Harbor Act of 1899,[18] the building of any structure, removal of any material, or the undertaking of any activity altering the condition of any navigable river or water of the United States is unlawful unless authorized by the Secretary of the Army on plans recommended by the Chief of Engineers.[19] By Section 4(f) of the OCSLA,[20] the Corps claims authority to prevent obstructions to navigation and alterations in the waters above the outer continental shelf without regard to the use to which the structure is put.[21] However, as noted above, this claim should be buttressed by legislative fiat. Additionally, if implantation of mooring equipment at the site of each OTEC station and any laying of plant to shore submarine cable will involve dredging of ocean bottom, a permit under Section 10 of the River and Harbor Act will be required.

The Corps generally issues dredging permits under Section 10 and under Section 404 of the Federal Water Pollution Control Act (FWPCA) of 1972.[22] Section 404 applies to discharges of dredged or fill material into the navigable waters at disposal sites specified by the administrator of the Environmental Protection Agency.[23] *Navigable waters* are defined in FWPCA as "waters of the United States, including the territorial seas."[24] Section 404 thus appears not to apply to discharges in the waters superjacent to the outer continental shelf. This shortcoming might be remedied by Congress when considering legislation establishing an OTEC plant siting program and regulation of construction and operation of OTEC plants.

Finally, Section 101 of the Marine Protection, Research and Sanctuaries Act of 1972[25] prohibits the transportation of any material,[26] from the United States or, in the case of a U.S. registered vessel, from any location for the purpose of dumping it into ocean waters, except in accordance with a permit issued under Section 103 by the Secretary of the Army. *Ocean waters* are defined in the Act as the open seas lying seaward of the baseline from which the territorial sea is measured.[27] Dredging and filling of the seabed in connection with installation of any required mooring equipment and laying of transmission cable for OTEC plants would be permitted by the Corps under this provision of the law, in addition to those sections discussed above. The administrator of the Environmental Protection Agency can veto the issuance of ocean dumping permits upon a finding that the dumping would create an adverse environmental effect on the quality of the receiving waters or their resources.[28]

Permit Procedure. A Corps permit for dredging, filling, and siting would be issued to a company proposing such activities in connection with the construction and operation of an OTEC plant as part of the one-stop federal authorization procedure recommended above for OTEC plants. In the absence of such a unified system, application would be submitted to the district engineer having jurisdiction over the district in which the plant is to be sited. Notice to the public regarding the proposed activity, preparation of a draft environmental-impact statement, public comments, a hearing, revision of the impact statement, and an announcement of decision would be similar to that described above for licensing under a single-authorization method.[29]

United States Coast Guard

The Coast Guard will have jurisdiction over OTEC plants in three general areas: (1) inspection and certification of plant manufacture or construction; (2) siting and emergency preparedness; and (3) operation.

General Statutory Authority. Congress has invested the Coast Guard under 14 U.S.C. §2 with *inter alia*, the enforcement of all applicable federal laws on and under the high seas and navigable waters of the United States; the administration of laws and regulations to promote the safety of life and property on and under the high seas and navigable waters; and the establishment and maintenance of aids to navigation for the promotion of safety on the high seas and waters subject to the jurisdiction of the United States.[30] Under 14 U.S.C. § 633 and 46 U.S.C. § 416, the Secretary of the cabinet department in which the Coast Guard is operating is empowered to promulgate regulations and orders deemed appropriate to carry out any provision of law applicable to the Coast Guard.[31]

Requirements Concerning Construction, Equipment, Measurement, and Documentation. The Coast Guard is required to inspect and certify merchant vessels documented under the laws of the United States. An OTEC plant, while differing in configuration and purpose from all other floating craft certified by the Coast Guard, would nonetheless be subject to traditional plan approval and inspection requirements applied to other U.S. flag vessels.

Hull Construction. An OTEC plant probably would be classified as a seagoing barge of capacity greater than 100 tons. Section 395(a) of Title 46 U.S. Code[32] requires the Coast Guard to inspect, before they are put into service and every two years thereafter, the hulls and equipment of seagoing barges of 100 gross tons or over that do not carry passengers to determine whether they are in conformity with the law and regulations of the Coast Guard. Thus OTEC plants would be required to meet the rules respecting construction of hull, decks, bulkheads, superstructure, railings, and crew accommodations.[33] Hull construction to the standards of the American Bureau of Shipping is required.[34] Impact and weather stability standards must be achieved.[35]

Equipment. Boilers, piping, and unfired pressure vessels of all vessels subject to inspection are similarly subject to inspection and certification of compliance.[36] The materials comprising boilers and unfired pressure vessels must be approved, tested, inspected, and stamped by the Coast Guard.[37] The Coast Guard is empowered to prescribe formulas, rules, and regulations for the design, material, construction, and operation of boilers, unfired pressure vessels, piping, valves, fittings, and other appurtenances for use on vessels subject to inspection.[38] The evaporator, condenser, and heat exchangers of an OTEC facility would probably be classified as "pressure vessels" by the Coast Guard and be subject to strict standards of design, construction, and testing as provided in 46 C.F.R. § 54. The Boiler and Pressure Vessel Code of the American Society of Mechanical Engineers (ASME), division 1, section VIII, is incorporated in Section 54 and must be followed.[39] Standards for class B, C, or R heat exchangers of the Tubular Exchanger Manufacturers Association are also incorporated and applicable.[40] Toughness test requirements of the American Society of Testing Materials (ASTM) obtain also.[41]

OTEC plant piping systems shall be designed, constructed, and inspected in accordance with the Code for Pressure Piping and Power Piping of the American National Standards Institute. Some portions of the ASME and ASTM codes apply also.[42] Plant electrical engineering is governed by Coast Guard regulations. Electrical systems and equipment, emergency lighting and power systems, and communications and alarms systems and equipment must be designed, installed, and operated according to Subchapter J of Title 46 C.F.R.[43]

Load Lines. U.S. flag vessels on voyages to or from foreign ports and coastwise voyages must be surveyed and issued load-line certificates to indicate the minimum freeboard to which each may be safely loaded. Load lines must be permanently and conspicuously marked and maintained on vessels.[44] The statute applies the requirements, however, to vessels on "voyages."[45] OTEC plants, when being towed initially to their operating sites and when returned to and from dry dock for overhaul, are seemingly not on "voyages" within the meaning of the statute. Moreover, they will not be loaded with cargo and hence would seem not to require survey for the purpose of establishing a load line. This area is one the Coast Guard must address prior to the application of its load-line regulations to OTEC facilities.[46]

Measurement and Documentation. Before an OTEC plant is documented as a U.S. flag vessel it must be admeasured by the Coast Guard in accordance with methods prescribed by Congress and implemented by the Secretary of Transportation.[47] Under 46 C.F.R. § 69.05-15 the marine document of admeasurement must show the date and place of construction, register length, breadth, depth, height, tonnage capacity, and other matters.

An OTEC plant may be documented under the laws of the United States when owned by a citizen of the United States.[48] However, since an OTEC station is not engaged in "trade," there would seem to be no requirement that it be documented as a U.S. flag vessel as long as it is registered.[49] Marine documents of registration, license or enrollment, and license are issued by the Coast Guard's officer in charge of marine inspection for the zone in which the vessel is "home-ported."[50]

Requirements Concerning Emergency Preparedness and Siting. (1) *Emergency Procedures.* The Secretary of Transportation is empowered to prescribe rules and regulations for vessel safety inspection and certification with regard to:

1. Lifesaving equipment (number, type, size, and capacity)
2. Firefighting equipment and precautionary measures guarding against fire (type of equipment, manning of equipment, drills to assure proper use of equipment)
3. Muster lists, prescribing special duties to be performed in emergencies
4. Ground tackle and hawsers
5. Bilge systems (designated capacity, manning, and inspection).[51]

OTEC plants will be required to meet Coast Guard standards relative to possession of lifesaving and firefighting equipment, marking of emergency evacuation routes, establishment of emergency power systems, manning of lifeboats, and so on.[52]

(2) *Siting.* The Secretary of Transportation is authorized to prescribe and enforce necessary and reasonable rules and regulations for the establishment, maintenance, and operation of lights and other signals on fixed and floating structures in or over waters subject to the jurisdiction of the United States and in the high seas for structures owned or operated by persons subject to the jurisdiction of the United States.[53] OTEC plants would be required, among other things, to be outfitted with aids to navigation, including lights and sounding devices, and to comply with orders requiring the establishment of a safety zone with buoys around the perimeter of a plant into which ships could not intrude. A 500-m safety zone is recommended (see Chapter 5).

Requirements Concerning Operation. (1) *Dangerous Cargo Regulation.* Under federal law no vessel on navigable waters, whether arriving or departing, or while in drydock, may carry or use except as fuel any explosives or dangerous articles or substances, except as permitted by regulations of the Coast Guard. "Explosives or dangerous articles or substances" includes flammable liquids and compressed gases.[54] Suggested designs of OTEC plants incorporate ammonia, Freon, or propane as a working medium. Although an OTEC plant will not be laden with these hazardous media while in navigable waters, arriving or departing port, or in dry dock, the Coast Guard, under the general police powers conferred in 14 U.S.C. § 2, could regulate the use of the substances beyond the territorial sea and prescribe the methods of design, construction, and testing of equipment that handles them. Under existing regulations, the presence on board of liquid flammable or combustible cargoes in bulk must be approved by the Coast Guard through endorsement on the vessel's certificate of inspection.[55] Extensive regulations exist prescribing standards for tank vessels carrying liquid flammable or combustible cargoes[56] and for other vessels carrying dangerous or explosive cargoes in bulk.[57] The Coast Guard must determine which portions of its rules will be applied to the working media of OTEC plants and then apply them uniformly.

(2) *Manning.* The Coast Guard has the authority to establish the minimum number of officers and crew an OTEC plant must have to ensure its safe operation, and it can require the vessel's certificate of inspection to reflect that determination.[58] Manning requirements for an OTEC vessel will be set by the officer in charge of marine inspection after inspection of the plant following construction. A variety of factors determines manning requirements, including size, type, and purpose of the vessel, as well as location.[59] Rules regarding minimum number of watches and overtime can be set also.[60] In setting manning requirements, the officer in charge would most likely provide different requirements for the vessel when navigating than when operating, and different crew qualifications for those personnel engaged in navigation as opposed to those in operation.

(3) *Sewage Treatment.* Section 312(h)(4) of the Federal Water Pollution Control Act prohibits a vessel subject to the standards and regulations promulgated under the Act (except a vessel not equipped with installed toilet facilities) from operating on the navigable waters of the United States unless the vessel is equipped with an operable marine sanitation device certified pursuant to the Act.[61] OTEC plants will be sited well beyond the 3-mi reach of U.S. "navigable waters." Accordingly, they will not be required to be equipped with marine sanitation devices for the treatment of sewage. As a matter of public policy, however, the owners or operators of the plants may wish to install such devices, or a requirement may be imposed by Congress in the enactment of legislation authorizing the siting and licensing of OTEC stations.

Approval Procedures. Coast Guard approval of plans for vessel construction, use of equipment, measurement, load lines, aids to navigation, use of dangerous substances, and manning would be issued to a person proposing to construct and operate an OTEC plant as part of the one-stop federal authorization procedure recommended above for OTEC plants. Under such a scheme, the lead agency, upon receipt of an application, would require (pursuant to a prior agreement with the Coast Guard) completion of CG form 3752, Application for Inspection of U.S. Vessel, or CG 986, Application for Inspection of Foreign Vessel. The completed form and construction plans are referred to the officer in charge of marine inspection for the district in which inspection and construction are to occur. The plans are reviewed by the officer in charge, with some review and approval delegated to the American Bureau of Shipping, and stamped approved (or denied, as the case may be). During and after construction, inspection is made and a certificate is then issued.[62]

Environmental Protection Agency

Questions Presented. Operation of an OTEC plant will result in the emission of seawater at different temperatures than when introduced into the plant, and thus the question of water pollution permits is raised. In the plant's evaporation water system, water warmed by the sun is skimmed from the ocean's surface, piped to a boiler, and used to evaporate a working fluid, such as ammonia, Freon, or propane, and then discharged into the ocean.[63] In the course of its trip through the plant, the water will be cooled and will accumulate metal particles from plant equipment. Chemicals may also be introduced into the stream to inhibit corrosion and purge marine growth. The water will be discharged many feet below its intake into a water mass of lower temperature than that from which it was originally taken.[64]

In the plant's condenser cooling water system, cold water is pumped in from an ocean stratum many feet below the surface, piped to a heat exchanger to condense the working medium from gas to liquid, and then discharged into the ocean. This water will be discharged at a higher temperature than when introduced into the plant, will contain metals and chemicals, and will be emitted some feet above its intake point into a water mass of higher temperature than that from which it was taken.[65]

Two legal issues exist in relation to normal operating discharges. Are plant emissions "pollutants" or "pollution" and thus controllable under the Federal Water Pollution Control Act or related laws? Does existing law cover discharges beyond the territorial sea?

FWPCA: Water Effluents as "Pollution." Section 301(a) of the Federal Water Pollution Control Act of 1972[66] proscribes all discharges of pollutants except in accordance with certain sections of the Act, among them Section 402. Under FWPCA, a *pollutant* is defined in Section 502(6) as "dredged spoil, solid waste, incinerator residue, sewage, garbage, sewage sludge, munitions, chemical waste, biological materials, radioactive materials, heat . . . and industrial, municipal, and agricultural waste discharged into water," and *pollution* is defined in Section 502(19) as "the man-made or man induced alteration of the chemical, physical, biological and radiological integrity of water."[67] There thus seems to be little doubt that OTEC plant discharges are "pollution" within the meaning and intent of the Act.

FWPCA: Exemption of Vessel beyond Territorial Sea. (1) *Discharge of a Pollutant.* Operating OTEC plants beyond the U.S. territorial sea clouds the question of federal authority to regulate plant discharges. The prohibition in Section 301(a) of FWPCA is modified by the permitting sections, 402 and 403. Section 403 prohibits "discharge[s] into the territorial sea, the waters of the contiguous zone, or the oceans"[68] except under a Section 402 permit and in compliance with guidelines promulgated by the administrator of the Environmental Protection Agency regarding nondegradation of the water quality and resources of the receiving water body.[69]

Section 402, National Pollutant Discharge Elimination System (NPDES), provides for the issuance of permits by the Environmental Protection Agency for the discharge of any pollutant.[70] Section 502(12) limits the definition of "discharge of a pollutant" as follows:

> (12) The term "discharge of a pollutant" and the term "discharge of pollutants" each means (A) any addition of any pollutant to navigable waters from any point source, (B) any addition of any pollutant to the waters of the continguous zone or the ocean from any point source other than a vessel or other floating craft.[71]

Point source is defined in the Act to include virtually every type of device capable of emitting a pollutant.[72] Discharges of pollutants from any point source into the waters of the contiguous zone and high seas require a permit under Sections 402 and 403, except that discharges from vessels and other floating craft are specifically exempted.

(2) *OTEC plants as Vessels.* Are OTEC plants, as presently conceived, vessels or floating craft? Current configurations show a large floating spar buoy, twin pontoons, or a floating platform, containing an entire central station power plant, moored to the seabed by a long anchor or floating freely.[73] *Vessel* in the law of admiralty "is applied to floating structures capable of transporting something over the water,"[74] notwithstanding the absence of a motive means of propulsion.[75] Although some courts have held that the watercraft must be practically capable of being used for transportation and must not be permanently located at one spot,[76] the general rule followed by the courts in construing the term *vessel* has been to give a very broad meaning to the term *vessel.*[77] An OTEC plant doubtless is a vessel within the meaning of Section 502(12) FWPCA.

Even assuming an OTEC plant is not a vessel, it is a watercraft not permanently affixed to the bottom and therefore could be classified as "other floating craft" as that term is used in Section 502(12).

Emanating from a vessel or floating craft, the effluent from the OTEC plant's evaporative-water and condenser-cooling-water systems would not be the "discharge of a pollutant" as that term is defined in Section 502(12). It follows, then, that OTEC plant discharges into waters beyond the U.S. territorial sea are not subject to the constraints of FWPCA.

Other Water Pollution Laws. OTEC plant discharges are not subject to other provisions of federal law either. The Marine Protection, Research and Sanctuaries Act of 1972 in Section 101 prohibits the transporting of "material" for the purpose of dumping it into waters of the contiguous zone or high seas from U.S. registered vessels or by any United States agency.[78] However, the definition of *material* does not include water or heat or other substance normally emitted by power plants.[79] Section 13 of the River and Harbor Act of 1899, as amended,[80] authorizes the Corps of Engineers to issue permits for the discharge of refuse into navigable waters. "Navigable waters" do not extend beyond the U.S. territorial sea, and the Corps of Engineers has never attempted to extend its Section 13 authority to include the waters above the outer continental shelf.

EPA: Anticipating OTEC Facilities. The coverage of the Federal Water Pollution Control Act can be extended by Congress to reach OTEC facilities operating beyond the U.S. territorial sea. (The Muskie bill now pending in the Senate would extend all U.S. pollution laws to activities in an area similar in size to an economic zone under consideration by the Third United Nations Conference on the Law of the Sea in the Informal Single Negotiating Text.) When faced with a similar question respecting deepwater ports, Congress responded by decreeing that:

> *the Constitution, laws, and treaties of the United States shall apply to a deepwater port licensed under this Act* and to activities connected, associated, or potentially interfering with the use or operation of any such port, *in the same manner as if such port were an area of exclusive Federal jurisdiction located within a State*. Nothing in this Act shall be construed to relieve, exempt, or immunize any person from any other requirement imposed by Federal law, regulation, or treaty. Deepwater ports licensed under this Act do not possess the status of islands and have no territorial seas of their own.[81] [Emphasis added]

Section 301 Effluent Limitations. Such an enactment vis-à-vis OTEC facilities would have the effect of applying the restrictions of FWPCA to the plant's evaporative water system discharges and condenser cooling water system discharges. A Section 402 NPDES permit issued by the administrator of the

Environmental Protection Agency would be required prior to operation. The issues considered in connection with the granting of a Section 402 permit are whether the discharges will comply with Section 301 of the Act (among other sections). Section 301 requires achievement of effluent limitations by July 1, 1977 for point sources reflecting the application of best practicable control technology currently available, and by July 1, 1983 for categories and classes of point sources reflecting the application of the best available technology relative to the control, reduction, or elimination of pollutants emitted by OTEC plants. Analysis of the effects of discharges on plankton, fish, shellfish, shorelines, and beaches; esthetic, recreation, and economic values; and the persistence and permanence of the effects of the discharges, among other matters, are all relevant considerations.[82]

Section 316(a) Exemption. EPA has published regulations governing the operation of steam electric power plants.[83] A brief examination of the provisions for information relative to chemical and thermal tolerances allowed is appropriate, although the regulations are inapplicable to OTEC plants.[84] For chemical components in power plant water discharges, limits are set for a number of pollutants, including total suspended solids, oil and grease, copper, iron, free available chlorine, and corrosion-inhibiting substances. Heat from the main condensers in new power plants may not be discharged from the cooling water into a natural water body,[85] and thus most new land-based plants will generally be required to install offstream cooling ponds or towers. However, the plant operator may obtain an exemption under Section 316(a) of FWPCA from the no-heat-discharge standard.[86] The exemption is obtainable upon demonstration that the standard as applied to the thermal component of the discharge is more stringent than necessary to "assure the protection and propagation of a balanced, indigenous population of shellfish, fish and wildlife in and on the body of water into which the discharge is to be made. . . ."[87]

OTEC plants sited in the ocean or the Gulf of Mexico seemingly would have no difficulty securing a Section 316(a) exemption from thermal-effluent limitations imposed under Section 301, since thermal discharges should have little impact on the ecology of the receiving water body.

Discharge Permit Procedures. An NPDES permit and a Section 316(a) exemption would be issued to a company or authority proposing to operate an OTEC plant by EPA as part of the one-stop federal authorization procedure recommended above for OTEC plants. The permit would contain conditions applicable to discharges, and constitute certification of compliance with FWPCA.

Current Discharge Permit Procedures. In the absence of such a unified system, application for an NDPES permit including a 316(a) exemption is submitted to the regional administrator of EPA in the region where the plant is to be sited.

Region IV, Atlanta, covers the Southeast, including Florida. Notice of the NPDES application is circulated to interested members of the public. The notice includes a tentative determination to issue or deny the permit and an announcement of an impending hearing. Hearings are required under Section 402 when requested by the applicant or by another person with significant public interest, or by the regional administrator at his own discretion, even when no request for one has been made.[88]

An environmental-impact statement would not have to be prepared by EPA in connection with an NDPES permit for OTEC power plant discharges. Section 511(c)(1) of FWPCA exempts all actions of the administrator from the NEPA statement requirement, except Section 402 permits for the "discharge of any pollutant by a new source" (and one other action not relevant to our discussion).[89] OTEC discharges are not the "discharge of a pollutant" under Section 502(12) because they emanate from a "vessel" or "floating craft" and because OTEC facilities are not "new sources" within the meaning of Section 306(a)(2) and (3) FWPCA.

Hazardous Substance Liability. Section 311(b)(3) of FWPCA makes unlawful the discharge of oil or hazardous substances into or upon the navigable waters or waters of the contiguous zone in harmful quantities as determined by the President.[90] Section 311(b)(2)(A) of FWPCA requires the administrator of EPA to:

> develop, promulgate, and revise as may be appropriate, regulations designating as hazardous substances, other than oil as defined in this section, such elements and compounds which, when discharged in any quantity into or upon the navigable waters of the United States or adjoining shorelines or the waters of the contiguous zone, present an imminent and substantial danger to the public health or welfare, including, but not limited to, fish, shellfish, wildlife, shorelines, and beaches.[91]

The administrator shall also include a determination as to whether any such designated hazardous substance can actually be removed.[92] As of this writing, final regulations delineating hazardous substances have not been published. When they are, they undoubtedly will include ammonia, and possibly Freon or propane—the three suggested working media in OTEC plants. If these substances are so designated, and if the designation includes a determination that when discharged into ocean waters, they cannot actually be removed, the owner or operator of a vessel or offshore facility discharging one of these substances will be liable to the United States for a penalty based on the toxicity, degradability, and dispersability of the substance, not exceeding $5000 or a penalty determined by the amounts discharged and the toxicity, etc., of the substance, not exceeding $5 million.[93] The statute imposes criminal liability (fine not exceeding $10,000 or one year's imprisonment) on the person in charge of a vessel or offshore facility for failure to notify the United States immediately of a hazardous-substance discharge[94] and, on the owner or operator, a civil penalty not exceeding $5000 for each offense for any discharge.[95] Additionally, the owner or operator of a vessel is liable to the United States for the costs of cleanup in an amount not to exceed $100 per gross ton of such vessel or $14 million, whichever is lesser. Absolute liability is imposed where the discharge was as a result of willful negligence or willful misconduct. The costs constitute a maritime lien on the vessel recoverable in an *in rem* action in federal court.[96]

No liability for the costs of cleanup for discharge of oil or a hazardous substance is imposed on an owner or operator who can prove that the discharge was caused by an act of God or war, negligence on the part of the United States, or act of a third party, irrespective of negligence.[97] Similarly, no liability for penalties for the discharge of a substance not amenable to removal is imposed when such proof is available.[98] Moreover, when an owner or operator proceeds to remove oil or a hazardous substance pursuant to regulations promulgated as a result of Section 311, he may recover cleanup costs against the United States upon showing that the discharge occurred as a result of an act of God or war, negligence on the part of the United States, or an act of a third party, irrespective of negligence.

Energy Research and Development Administration

Title I of the Energy Reorganization Act of 1974 created the Energy Research and Development Administration, with authority to conduct research and development into, among other things, solar energy.[99] Included in the authority is the responsibility to encourage and conduct programs demonstrating the commercial feasibility and practical application of the extraction, conversion, storage, transmission, and utilization of solar energy.[100] Accordingly, OTEC technology would be eligible for continuing support during any experimental and demonstration phases through subsidies and awards made by ERDA.

Federal Energy Administration

The Federal Energy Administration is responsible by law for the development and implementation of projects and programs for the production and use of energy and the development of energy sources in the United States, as well as for devising and effectuating policies and programs to expedite the construction and licensing of U.S. energy facilities.[101] The oversight or management of OTEC programs by FEA (following basic development by ERDA) could be an important means of ensuring the expeditious and full development of this energy source.

Federal Power Commission

The Federal Power Commission (FPC) under current law would have jurisdiction over the interstate transmission and sale, including the setting of rates, of the electric power generated by OTEC plants, as well as other aspects of wholesale transactions in electric power.[102] The FPC has the authority to establish regional districts for the voluntary interconnection and coordination of facilities for the generation, transmission, and sale of electricity in order to assure an abundant supply of economical electric power while properly utilizing and safeguarding natural resources.[103] The FPC would serve as coordinating agency for the proper and efficient allocation and use of the power supplied from an OTEC plant in an interstate coastal grid.

Some have projected employment of an OTEC station as a supplier of electricity to an on-site electro-process industry, such as metal refining or reduction, hydrogen production, or liquid-fuel production facilities. If fuel were then piped to shore, the FPC could be vested by Congress with jurisdiction to regulate the sale of the fuel in commerce.

If any OTEC plant is sited at the border of the United States or connects to facilities sited at such border, the construction, operation, and maintenance of the plant or facility must be licensed by the Federal Power Commission.[104] If an OTEC facility is to be used to generate electric energy for transmission to a foreign country, the operator must first obtain an order of the FPC authorizing it to do so.[105]

Department of Defense

Congress has delegated to the President broad authority to restrict the use of the seabed, waters, and air space of the territorial sea, contiguous zone, and outer continental shelf in the interest of national defense.[106] Pursuant to that authority the President has established numerous defensive sea areas adjacent to the coast of the United States in which the siting of fixed structures and some traffic is prohibited. The Congressional authority has been upheld as within the constitutional power of providing for the national defense, and the delegation and exercise thereof by the President has been held proper to effectuate that end.[107]

Plans to site OTEC facilities in any defensive sea area will require approval of the Department of Defense and probably only upon proper establishment of national need. Clearance should be sought from the Deputy Assistant Secretary of Defense for Installations and Housing. It should be noted that the Department of Defense and Bureau of Land Management of the Department of the Interior have cooperated successfully in the establishment of modes of conduct required of operators of offshore drilling rigs in military areas on the outer continental shelf.[108]

Department of Commerce

National Oceanic and Atmospheric Administration (NOAA). Among its responsibilities, NOAA has authority to ensure the proper management, use, and conservation of the ocean and its living resources, including commercial fisheries conservation, prevention of pollution, shellfish sanitation, protection of environmentally critical areas, and the safe transportation of hazardous materials. It has no direct licensing jurisdiction over artificial structures such as OTEC plants. Nevertheless, its comments on environmental impact statements prepared in connection with the federal action allowing the construction and operation of the facilities will be important, since the licensed activity will have significant impact on the ocean environment. NOAA's jurisdiction over state coastal zone management plans is relevant to the question of siting of OTEC plants. This process is well described in Chapter 10.

Maritime Administration. During war or during national emergency, as declared by presidential proclamation, no vessel (including a vessel under construction) owned by a citizen of the United States may be placed under foreign registry or flag, or sold, mortgaged, loaned, or in any way transferred to any person not a citizen of the United States without the approval of the Secretary of Commerce. No vessel constructed in the United States may be documented under foreign law without similar approval.[109] The Maritime Administration has published regulations setting forth criteria for approval. In evaluating applications, consideration is given to type, size, acceptability of foreign country of registry and buyer, and the "need to retain the vessel under U.S. flag or ownership for the purposes of national defense, maintenance of an adequate merchant marine, foreign policy of the United States, and the national interest."[110]

Department of the Interior

Bureau of Sport Fisheries and Wildlife (BSF&W). BSF&W has basic jurisdiction over the conservation, development, and management of domestic fish and wildlife resources and their environments, inland and offshore. In the case of OTEC plants, its comments on the environmental-impact statements issued in connection with the project authorization would be significant. Thus, while possessing no permitting authority in connection with OTEC activities, it will have a persuasive voice in the assessment of environmental impact under NEPA.

U.S. Geological Survey (USGS). The U.S. Geological Survey in the Department of Interior holds jurisdiction over federal lands insofar as they contain leasable minerals. In addition to classifying said lands, it supervises the activities of private concerns engaged in the extraction of minerals from public lands so as to maximize productive capacity and prevent waste and pollution. Since it is responsible for the maximum utilization of mineral resources USGS should be consulted by the operator of any OTEC facility prior to the selection of an operating site so as to preclude use of lands usable for the production of minerals (see discussion in Chapter 5).

Department of Labor: Occupational Safety and Health Administration

Under the Williams-Steiger Occupational Safety and Health Act of 1970[111] the Department of Labor is required to promulgate by rule as an occupational safety or health standard any national consensus standard and any established federal standard.[112] The standards as published apply to employments performed in the United States and, among other places, in "Outer Continental Shelf lands defined in the Outer Continental Shelf Lands Act."[113] Section 2(a) of that Act defines the lands as:

> . . . all submerged lands lying seaward and outside of the area of lands beneath navigable waters as defined in Section 2 of the Submerged Lands Act . . . and of which the subsoil and seabed appertain to the United States and are subject to its jurisdiction and control. . . .[114]

As a coastal state the United States has jurisdiction and control over lands lying seaward of the territorial sea for limited purposes, such as customs, sanitation, immigration, navigation, fishing, national defense, exploration, and exploitation of minerals.[115] Accordingly, it would seem that the Occupational Safety and Health (OS&H) Act would apply to employment of persons on an OTEC facility in waters superjacent to the outer continental shelf.

The OS&H Administration and the Coast Guard have worked unsuccessfully during the course of the past two years on a memorandum of understanding relative to the establishment and enforcement of standards for vessels and artificial structures operating in navigable waters and on the high seas. The agencies cannot agree on the sharing of responsibility. The OS&H Administration claims that the Act gives it jurisdiction to establish and enforce standards for vessels and structures unless another agency has preempted the field. The Coast Guard maintains that it is in the field and is developing standards. Meanwhile, the OS&H Administration is looking on and objecting. An earlier plan suggested that the agencies allocate matters relating to purely maritime aspects to the Coast Guard and aspects of the workplace not peculiar to a vessel or marine structure to the Occupational Safety and Health Administration.

State Jurisdiction

General

OTEC plants, under present siting schemes, are to be located beyond offshore sovereign lands owned by the states by virtue of the conveyance by the federal government in the Submerged Lands Act.[116] If OTEC plants are used to generate electricity for transmission to shore, however, transmission cables will traverse state lands. The laying and burying of the cables will result in the dredging and filling of seabed and beaches owned by the states and the establishment of permanent obstructions on state lands. An onshore substation and switchyard must be constructed for reception of the electrical power and distribution to other facilities of the coastal utility. A dock and warehouse must be constructed on shore (and may require dredging) to service the OTEC plant. Perhaps even a heliport will be needed on shore for expeditious transfer of crew and maintenance employees to the plant. The distribution and sale of power generated by the plants will be in the adjacent state or states. In summary, the construction and operation of service facilities and transmission lines, the operation of OTEC plants themselves, and the distribution and sale of power generated by the plants will have a significant impact on particular coastal states, their resources, lands, waters, and economy. Accordingly, the state will have a vital interest in the regulation of most facets of OTEC facilities.

The states have no title to; proprietary interest in; or powers, dominion, or control over lands and adjacent waters and airspace beyond the territorial sea, and therefore would have no direct authority to license an OTEC facility (exclusive of transmission lines and onshore facilities). Nevertheless, any federal regulatory regime allowing the siting of OTEC plants in waters of the contiguous zone must recognize a valid state interest in the construction, siting, and operation of the plants and incorporate such interest in statutory and administrative authorization. Recognition of this interest was made by Congress when, in enacting the Deepwater Port Act (DPA) of 1974,[117] it granted to any state located within 15 mi of any proposed port, or any state within whose borders a pipeline from a proposed port is to be constructed, the right to disapprove licensing of the facility. A similar voice in the decisionmaking process relative to Atlantic leases of offshore oil deposits is currently being debated.

Virtually every coastal state in the United States has regulations controlling the use of offshore lands and waters for other than recreational and normal commercial purposes. Since current siting requirements dictate that OTEC plants be located in subtropical regions, it would seem pertinent to examine a Southern coastal state's rules appropriate to an artificial structure moored in waters on the adjacent outer continental shelf. Florida would appear to be an appropriate example.

Overall Project Review by States

Power Plant Siting Acts. Some coastal states in recent years have enacted "one-stop" siting laws for power plant construction. Generally, the laws provide for utility companies to file with the state a 10-year site plan estimating power generating needs and the general location of proposed plant sites. The state performs a preliminary study of the plan and assesses the need for power, environmental impact, anticipated growth, conformance with comprehensive land-use plans, and other matters. A utility may apply to a "lead agency" of the state for certification of any site, transmission lines, and auxiliary facilities in its 10-year plan. The application undergoes a comprehensive review by appropriate state agencies, an environmental-impact statement is written, a public hearing is held, and a decision is rendered. Once certified, the state is bound as to approval of site, plant, and transmission lines; the site may be used by the utility for plant construction and operation and no additional state permits or authorizations are required. OTEC facilities could be sited, in part, under such one-stop siting programs. Any transmission lines and shore-based dock, warehouse, and substation would be eligible. Any electric utility system planning the construction of an OTEC facility would be required to include the project in its 10-year plan. In Florida the Department of Environmental Regulation serves as certifying agency with the Division of State Planning performing studies of 10-year plans and the Public Service Commission providing recommendations as to the need for power.[118]

Coastal Zone Management Plans. In many coastal states, state agencies are currently compiling inventories and charting the resources of the coastal zone preparatory to writing comprehensive coastal zone land-use plans. Where one-stop siting laws have not been enacted, any company or authority proposing the construction of OTEC facilities would have to seek approval of, or (at a minimum) inform, coastal zone or land-use planning agencies of those aspects of the proposed activity which impact on state lands (transmission lines and shore facilities, for example). In Florida the Division of Resource Management of the Department of Natural Resources, which has statutory authority to "develop a comprehensive state plan for the protection, development and zoning of the coastal zone,"[119] serves as an advisory organ to the governor and cabinet in matters related to conservation, preservation, and development of the Florida coastal zone. The Division should be consulted in connection with proposed siting of OTEC facilities. Thomas Stoel discusses the state-federal relationships under the Federal Coastal Zone Management Act in Chapter 10.[120]

Individual Facility Component Review by States

In the absence of one-stop site certification laws, the transmission lines and auxiliary shore-based facilities of an OTEC plant would undergo separate permitting by individual state agencies.

Construction of Transmission Cables and Land Support Facilities. The dredging and filling operations connected with the laying of submarine cable and land-support facilities (if required) are generally done only under permit. In

Florida the Board of Trustees of the Internal Improvement Trust Fund holds title to all state offshore lands,[121] but the issuance of permits for the dredging and filling therein is done by the Department of Environmental Regulation.[122] Review of such permit applications would also generally be conducted by state departments vested with jurisdiction over fish and waterfowl, natural resources, esthetics, shoreline preservation, etc. In Florida these reviews are performed by the Department of Natural Resources.[123] Also in Florida, as in most states, the permit applicant must obtain a certificate from the Department of Environmental Regulation that the dredging and filling will be in compliance with applicable effluent limitations and water-quality standards.[124] Additionally, this certificate must be filed in connection with the application to the Corps of Engineers for a dredge-and-fill and siting permit for the project.[125] Moreover, most states require approval of local municipalities or counties before entertaining applications for dredge-and-fill permits.[126] Some states prohibit construction below beach setback lines but allow variances upon a proper showing of public need.[127] Local building permits for cable and support facility construction must also be secured.

If a heliport is planned for the land-support facilities, a license to construct and operate is generally required by state authorities. Minimum standards of size and shape must be met.[128]

Cable and Support-Facility Rights-of-Way. The laying of transmission cable in offshore lands will require a permanent easement from the state agency vested with permitting authority. In Florida this is the Department of Environmental Regulation.[129] Inside the mean high-water mark, an easement must be secured from the upland owner.

If support facilities are sited on public lands, grants, leases, or easements from the appropriate state agency must be secured.

Conclusions

From this discussion the following observations might be offered:

(1) No private right or power exists under current U.S. law enabling or authorizing the siting or operating of an OTEC plant beyond the U.S. territorial sea. Congress must therefore enact legislation legitimizing OTEC siting in areas beyond the territorial sea.

(2) Once Congress asserts national jurisdiction to site and operate OTEC plants beyond the 3-mi limit, a streamlined federal regulatory regime should be established for licensing of the facilities. Several federal agencies exist as candidates for the licensing role.

(3) The states have a legitimate economic and environmental interest in OTEC power stations and should be given a role in the decisions connected with their siting, operation, and product distribution.

Notes

1. *See* Deepwater Port Act of 1974, 88 Stat. 2126 (1975); 33 U.S.C.A. § 1501 (Supp. 1975); 40 *Fed. Reg.* 52,539 (1975).

2. 42 U.S.C. § 2133 (Supp. 1975); 10 C.F.R. Part 50 (1974).

3. 33 U.S.C. §§ 401, 403 (Supp. 1975); 33 C.F.R. § 209.120 (1974); 16 U.S.C. § 797(1) (Supp. 1975); 18 C.F.R. Part 4 (1975).

4. 16 U.S.C. § 824a(e) (Supp. 1975); 18 C.F.R. Part 32 (1975).

5. *Supra* note 1.

6. The Senate Government Operations Committee now holds most power respecting energy planning and will have a major voice in shaping any federal OTEC program. Other important bodies are the Senate Interior, Commerce, and Banking and the House Interior, Commerce, Ways and Means, and Merchant Marine committees, each with jurisdiction over some aspect of energy or energy sources.

7. Deepwater Port Act of 1974, § 5(e)(2), *supra* note 1.

8. 42 U.S.C. § 4332 (Supp. 1975).

9. 16 U.S.C. § 1456(c)(3) (Supp. 1975).

10. *Cf.* Corps of Engineers Regulations on Dredge and Fill, 33 C.F.R. § 209.120(j)(l)(ix)(a) (1974).

11. Deepwater Port Act of 1974, § 5(i)(2), *supra* note 1.

12. Id.

13. 43 U.S.C. §§ 1333-1343 (Supp. 1975). Sec. 4(a)(1) and (2) of the Act [43 U.S.C. § 1333(a)(1) and (2)] provide:

Sec. 4. Laws Applicable to Outer Continental Shelf.–(a)(1) The Constitution and laws and civil and political jurisdiction of the United States are hereby extended to the subsoil and seabed of the outer Continental Shelf and to all artificial islands and fixed structures which may be erected thereon for the purpose of exploring for, developing, removing, and transporting resources therefrom, to the same extent as if the outer Continental Shelf were an area of exclusive Federal jurisdiction located within a State: *Provided, however,* That mineral leases on the outer Continental Shelf shall be maintained or issued only under the provisions of this Act.

(2) To the extent that they are applicable and not inconsistent with this Act or with other Federal laws and regulations of the Secretary now in effect or hereafter adopted, the civil and criminal laws of each adjacent State as of the effective date of this Act are hereby declared to be the law of the United States for that portion of the subsoil and seabed of the outer Continental Shelf, and artificial islands and fixed structures erected thereon, which would be within the area of the State if its boundaries were extended seaward to the outer margin of the outer Continental Shelf, and the President shall determine and publish in the Federal Register such projected lines extending seaward and defining each such

area. All of such applicable laws shall be administered and enforced by the appropriate officers and courts of the United States. State taxation laws shall not apply to the outer Continental Shelf.

14. *See* 40 *Fed. Reg.* 31,319 (July 25, 1975), amending 33 C.F.R. § 209.120(b)(2) (1974).

15. 43 U.S.C. § 1333(f) (Supp. 1975). Sec. 1333(f) provides:

(f) The authority of the Secretary of the Army to prevent obstruction to navigation in the navigable waters of the United States is hereby extended to artificial islands and fixed structures located on the outer Continental Shelf.

16. A leading case, U.S. v. Ray, 423 F.2d 16 (5th Cir. 1970), interpreting the extent of jurisdiction of the United States over structures not related to mineral exploration or extraction is of little help. Although upholding the federal government's authority to enjoin construction of an artificial island on coral reefs by private interests beyond the territorial sea, the court did not declare that the United States had plenary authority to regulate nonextractive uses.

17. *Supra* note 1.

18. 33 U.S.C. § 402 (Supp. 1975).

19. The Secretary of the Army has delegated his permitting authority to the Chief of Engineers for all authorizations for work in navigable waters, dredging and filling, and ocean dumping. *See* 33 C.F.R. § 209.120(p) (1974).

20. *Supra* note 14.

21. *Supra* note 13.

22. 33 U.S.C. § 1344 (Supp. 1975). Section 404(a) provides:

Sec. 404(a). The Secretary of the Army, acting through the Chief of Engineers, may issue permits, after notice and opportunity for public hearings for the discharge of dredged or fill material into the navigable waters at specified disposal sites.

23. The EPA regulations are in 40 C.F.R. Part 227 (1974).

24. FWPCA, § 502(7); 33 U.S.C.A. § 1362 (Supp. 1975).

25. 33 U.S.C.A. § 1413 (Supp. 1975).

26. Sec. 3(c) of the Act defines *material* as "matter of any kind or description, including but not limited to, dredged material. . . ." 33 U.S.C.A. § 1402 (Supp. 1975).

27. Marine Protection, Research, and Sanctuaries Act of 1972, § 3(b); 33 U.S.C. § 1402 (Supp. 1975).

28. Marine Protection, Research, and Sanctuaries Act of 1972, § 102(c); 33 U.S.C. § 1413(c).

29. *See* Corps Regulations on Dredge and Fill, 33 C.F.R. § 209.120 (1974), as amended by 40 *Fed. Reg.* 31,319 (July 25, 1975).

30. 14 U.S.C. § 2 (Supp. 1975).

31. 14 U.S.C. § 633 (Supp. 1975); 46 U.S.C. § 416 (Supp. 1975).

32. 70 Stat. 225 (1956).

33. 46 C.F.R. Part 92 (1974).

34. 46 C.F.R. § 92.01-10 (1974).

35. 46 C.F.R. § 93.07 (1974).

36. 46 U.S.C. § 392(b) and (c) (Supp. 1975).

37. 46 U.S.C. § 406 (Supp. 1975). Secs. 408, 409, and 412 of Title 46 prescribe additional requirements for boiler plates and the testing and inspection thereof of vessels subject to inspection. Secs. 407, 410, and 413 prohibit or provide sanctions for anyone using uninspected or defective equipment or counterfeit stamps or obstructing safety valves of a boiler.

38. 46 U.S.C.A. § 411 (Supp. 1975).

39. 46 C.F.R. § 54.01-1 (1974).

40. 46 C.F.R. § 54.01-2 (1974).

41. 46 C.F.R. § 54.01-3 (1974).

42. 46 C.F.R. §§ 56.01-5 and 6 (1974).

43. 46 C.F.R. Parts 110-113 (1974).

44. 46 U.S.C.A. § 86(c) (Supp. 1975).

45. Under 46 U.S.C.A. § 86b(a) load-line certificates are required of U.S. vessels arriving at or departing from any U.S. port or place within the jurisdiction of the United States from or for a foreign port. Under 46 U.S.C.A. § 88 load-line requirements are also applicable to vessels of 50 gross tons or over loading at or proceeding to sea from any port or place within the United States or its possessions for a coastwise voyage by sea. A "coastwise voyage by sea" is a voyage in which a vessel proceeds from one port or place in the United States or its possessions and passes outside the line dividing inland waters from high seas.

46. 46 C.F.R. Part 42 (1974) contains the Coast Guard requirements respecting load lines. The commandant may grant an exemption to vessels or categories of vessels from the requirements. 46 C.F.R. § 42.03-30.

47. 46 U.S.C. § 71 (Supp. 1975). Secs. 74, 75, and 77 prescribe the methods for admeasurement.

48. 46 U.S.C. §§ 11, 13, and 14 (Supp. 1975).

49. 46 U.S.C. § 319 (Supp. 1975). *Also see* 46 C.F.R. §§ 67.01-13 and 67.07-11 (1974). However, the U.S. Maritime Administration requires vessels constructed in American shipyards to be registered as U.S. flag vessels.

50. 46 C.F.R. § 67.07-5 (1974).

51. 46 U.S.C.A. § 481(a) (Supp. 1975).

52. *See* 46 C.F.R. Parts 94, 95, 97 (1974).

53. 14 U.S.C.A. § 85 (Supp. 1975).

54. 46 U.S.C. § 170 (Supp. 1975).

55. 46 C.F.R. § 30.10-49 (1974).

56. 46 C.F.R. Parts 30-40 (1974).

57. 46 C.F.R. Part 146 (1974).

58. 46 U.S.C. § 222 (Supp. 1975).

59. 46 U.S.C. §§ 157.15-1 and .20-15 to .20-60 (1974). *Also see* 46 C.F.R. § 31.15-1 (1974) for tank vessels.

60. 46 C.F.R. § 157.20-5 and .20-10 (1974).

61. 33 U.S.C. § 1322(h)(4) (Supp. 1975).

62. *See*, generally, 46 C.F.R. Part 2 (1974). For cargo and miscellaneous vessels, *see* Part 91 and for tank vessels, *see* Part 31.

63. *See* Chapter 1.

64. Id.

65. Id.

66. 33 U.S.C.A. § 1311(a) (Supp. 1975). Section 401(a) provides:

> Except as in compliance with this section and sections 302, 306, 307, 318, 402, and 404 of this Act, the discharge of any pollutant by any person shall be unlawful.

67. 33 U.S.C. §§ 1362(6) and (19) (Supp. 1975), respectively.

68. FWPCA, § 403(a); 33 U.S.C.A. § 1343(a). Subsection (a) reads in full as follows:

> Sec. 403(a) No permit under section 402 of this Act for a discharge into the territorial sea, the waters of the contiguous zone, or the oceans shall be issued, after promulgation of guidelines established under subsection (c) of this section, except in compliance with such guidelines. Prior to the promulgation of such guidelines, a permit may be issued under such section 402 if the Administrator determines it to be in the public interest.

69. FWPCA, § 403(c); 33 U.S.C.A. § 1343(c) reads in full as follows:

> (c)(1) The Administrator shall, within one hundred and eighty days after enactment of this Act (and from time to time thereafter), promulgate guidelines for determining the degradation of the waters of the territorial seas, the contiguous zone, and the oceans, which shall include:
>
> (A) the effect of disposal of pollutants on human health or welfare, including but not limited to plankton, fish, shellfish, wildlife, shorelines, and beaches;
>
> (B) the effect of disposal of pollutants on marine life including the transfer, concentration, and dispersal of pollutants or their byproducts through biological, physical, and chemical processes; changes in marine ecosystem diversity, productivity, and stability; and species and community population changes;

(C) the effect of disposal, of pollutants on esthetic, recreation, and economic values;

(D) the persistence and permanence of the effects of disposal of pollutants;

(E) the effect of the disposal at varying rates, of particular volumes and concentrations of pollutants;

(F) other possible locations and methods of disposal or recycling of pollutants including land-based alternatives; and

(G) the effect on alternate uses of the oceans, such as mineral exploitation and scientific study.

(2) In any event where insufficient information exists on any proposed discharge to make a reasonable judgment on any of the guidelines established pursuant to this subsection no permit shall be issued under section 402 of this Act.

70. FWPCA, Sec. 402 (a)(1); 33 U.S.C. § 1342, provides:

Sec. 402(a)(1). Except as provided in sections 318 and 404 of this Act, the Administrator may, after opportunity for public hearing, issue a permit for the discharge of any pollutant, or combination of pollutants, notwithstanding section 301(a), upon condition that such discharge will meet either all applicable requirements under sections 301, 302, 306, 307, 308, and 403 of this Act, or prior to the taking of necessary implementing actions relating to all such requirements, such conditions as the Administrator determines are necessary to carry out the provisions of this Act.

71. 33 U.S.C.A. § 3162(12). Subsection 502(12)(A) is not pertinent to regulation of OTEC plants since it is restricted to "navigable waters" and the definition of navigable waters under the Act does not include waters beyond the U.S. territorial sea. § 502(7) and (8); 33 U.S.C. § 1362(7) and (8).

72. Sec. 402(14); 33 U.S.C. § 1362(14), defines *point source*:

(14) The term "point source" means any discernible, confined and discrete conveyance, including but not limited to any pipe, ditch, channel, tunnel, conduit, well, discrete fissure, container, rolling stock, concentrated animal feeding operation, or vessel or other floating craft, from which pollutants are or may be discharged.

73. *See* Chapter 1.

74. Gilmore and Black, *Admiralty* 30.

75. Pleason v. Gulfport Shipbuilding Corp., 221 F.2d 621 (5th Cir. 1955), Rogers v. A Scow Without a Name, 80 F.736 (1897).

76. Evansville & Bowling Green Packet Co. v. Chero Cola Bottling Co., 271 U.S. 19 (1926), *The Quarterboat No. 130*, 19 F.S. 419 (1937). *Accord*, Cope v. Vallette Dry-Dock Co., 119 U.S. 625 (1887). Citations to the authorities in notes 74-76 were provided by the Office of the General Counsel, Environmental Protection Agency.

77. The Statutory definition of vessel is pertinent here, too. 1 U.S.C. § 3 (Supp. 1975) defines *vessel* as

every description of watercraft or other contrivance used or capable of being used as a means of transportation on water.

FWPCA, § 312(a); 33 U.S.C. § 1322(a) borrowed this traditional definition but restricts it to the subject of whether watercraft are subject to the prohibition of discharge of sewage overboard.

78. *Supra* note 25.

79. Marine Protection, Research and Sanctuaries Act of 1972, § 3(c); 33 U.S.C.A. § 1402(c), provides:

(c) "Material" means matter of any kind or description, including, but not limited to, dredged materials, solid waste, incinerator residue, garbage, sewage, sewage sludge, munitions, radiological, chemical, and biological warfare agents, radioactive materials, chemicals, biological and laboratory waste, wreck or discarded equipment, rock, sand, excavation debris, and industrial, municipal, agricultural, and other waste; but such term does not mean sewage from vessels within the meaning of section 312 of the Federal Water Pollution Control Act, as amended (33 U.S.C. 1332). Oil within the meaning of section 311 of the Federal Water Pollution Control Act, as amended (33 U.S.C. 1321), shall be included only to the extent that such oil is taken on board a vessel or aircraft for the purpose of dumping.

80. 33 U.S.C.A. § 407 (Supp. 1975).

81. Deepwater Port Act of 1974, § 19(a)(1); Pub. L. 93-627, 88 Stat. 2131 (1975).

82. FWPCA, § 403; *Supra* note 68, contains a pertinent listing of matters requiring evaluation in connection with discharges into ocean waters beyond the territorial sea.

83. 40 C.F.R. Part 423 (1974).

84. The steam electric power plant regulations were published as required by FWPCA, § 306(b), and apply to "new sources." *New source* under Section 306(a) is defined as:

any source, the construction of which is commenced after the publication of proposed regulations prescribing a standard of performance under this section which will be applicable to such source, if such standard is thereafter promulgated in accordance with this section.

Source under Sec. 306(a) is defined as:

any building, structure, facility, or installation from which there is or may be the discharge of pollutants.

These definitions do not include the all-encompassing language of "point source" found in the general definitions of the statute, § 502(14) (*supra* note 72). Specifically, they do not include the term *vessel or other floating craft*, which, we have seen, is what OTEC facilities doubtless are.

Moreover, the regulations themselves limit their applicability:

> to discharges resulting from the operation of a generating unit by an establishment primarily engaged in the generation of electricity for distribution and sale which results primarily from a process utilizing fossil-type fuel (coal, oil or gas) or nuclear fuel in conjunction with a thermal cycle employing the steam-water system as the thermodynamic medium.

40 C.F.R. § 423.10. Suggested designs of an OTEC plant utilize Freon, ammonia, or propane as a working medium and not steam.

85. 40 C.F.R. § 423.13(1) (1974).

86. 40 C.F.R. Part 122 (1974).

87. FWPCA, § 316(a); 33 U.S.C.A. § 1326(a) (Supp. 1975).

88. 40 C.F.R. § 125.32(e) (1974).

89. 33 U.S.C. § 1371(c)(1) (Supp. 1975). Subsection 511(c)(1) states in full:

> Except for the provision of Federal financial assistance for the purpose of assisting the construction of publicly owned treatment works as authorized by section 201 of this Act, and the issuance of a permit under section 402 of this Act for the discharge of any pollutant by a new source as defined in section 306 of this Act, no action of the Administrator taken pursuant to this Act shall be deemed a major Federal action significantly affecting the quality of the human environment within the meaning of the National Environmental Policy Act of 1969. . . .

90. 33 U.S.C. § 1321(b)(3) (Supp. 1975).

91. 33 U.S.C. § 1321(b)(2)(A).

92. FWPCA, § 311(b)(2)(B); 33 U.S.C. § 1321(b)(2)(B).

93. Id., §§ (i) and (iii).

94. FWPCA, § 311(b)(5); 33 U.S.C. § 1321(b)(5).

95. FWPCA, § 311(b)(6); 33 U.S.C. § 1321(b)(6).

96. FWPCA, § 311(f)(1); 33 U.S.C. § 1321(f)(1).

97. Id.

98. *Supra* note 90.

99. Energy Reorganization Act of 1974, § 103; Pub. L. 93-438, 88 Stat. 1233 (1974).

100. Id.

101. Federal Energy Administration Act of 1974, Pub. L. 93-275, 88 Stat. 98, 15 U.S.C.A. § 764 (Supp. 1975).

102. Federal Power Act, as amended, 49 Stat. 838 (1935), 16 U.S.C. §§ 824d, 824e (Supp. 1975).

103. Id., § 824a(a).

104. 18 C.F.R. Part 32 (1974).

105. 16 U.S.C. §§ 824d, 824e. *See* 18 C.F.R. Part 32 (1974) for procedures for obtaining an order.

106. 18 U.S.C. § 2152 (Supp. 1975).

107. Feliciano v. United States, 297 F.Supp. 1356 (D.C.P.R. 1969); Perko v. United States, 204 F.2d 446 (8th Cir. 1953), *cert. denied*, 346 U.S. 832, 74 S.Ct. 48 (1953).

108. 43 C.F.R. Part 3300.

109. Shipping Act of 1916, 39 Stat. 728; 46 U.S.C. § 835 (Supp. 1975).

110. 46 C.F.R. § 221.7, App. I(3).

111. 84 Stat. 1593 (1970); 29 U.S.C. § 655 (Supp. 1975).

112. A *national consensus standard* is a directive requiring the use of methods reasonably necessary to provide a safe or healthful place of employment which has been promulgated by a national standards-producing organization after study and has been designated as such by the Assistant Secretary of Labor for Occupational Safety and Health. An *established federal standard* is a similar directive established by an agency of the federal government and in effect on April 28, 1971 or contained in any Act of Congress in force on the date of passage of the Williams-Steiger Act. *See* 29 C.F.R. § 1910.1(f)-(h) (1974).

113. 29 C.F.R. § 1910.05(a) (1974).

114. 43 U.S.C. § 1331(a) (Supp. 1975).

115. Convention on the High Seas (1958), 13 U.S.T. 2312 (1962), T.I.A.S. No. 5200; convention on the Continental Shelf (1958), 15 U.S.T. 471 (1964), T.I.A.S. no. 5578, Article 2. *Also see* Chapter 3.

116. 67 Stat. 29 (1953), 43 U.S.C. §§ 1301-1315 (Supp. 1975).

117. *Supra* note 1. For a Supreme Court decision in this regard, *see* Askew v. American Waterways Operators, 411 U.S. 325 (1973).

118. Ch. 403.50 F.S. (Supp. 1974).

119. § 370.0211(4)(d) F.S. (Supp. 1974), as amended by Florida Environmental Reorganization Act of 1975 (hereafter Env. R. Act), Ch. 75-22 Laws of Florida.

120. *See* Chapter 10.

121. § 253.12 F.S. (Supp. 1974).

122. § 253.123 F.S. (Supp. 1974), as amended by Env. R. Act.

123. § 253.1241 F.S. (Supp. 1974), as amended by Env. R. Act and Ch. 75-153 Laws of Florida.

124. Ch. 403 F.S. (Supp. 1974), as amended by Env. R. Act. *See* Ch. 17-3 Fla. Adm. Code for Rules of the Department.

125. FWPCA, §§ 401(a)(1), 404; 33 U.S.C. §§ 1341, 1344 (Supp. 1975); River and Harbor Act of 1899, § 10; 33 U.S.C. § 402 (Supp. 1975).

126. *Cf.* § 253.124(3) F.S. (Supp. 1974), as amended by Env. R. Act.
127. *Cf.* § 161.053 F.S. (Supp. 1974).
128. *See* § 330.29 F.S. (Supp. 1974); § 14-60.06 Fla. Adm. Code (1974).
129. *Supra* notes 122 and 123.

10 Ocean Thermal Energy Conversion: Domestic Environmental Aspects

Thomas B. Stoel, Jr.

Environmental Impacts

The ocean thermal energy conversion plants referred to in this discussion of domestic environmental aspects are assumed to be among the four design configurations described in Chapter 1. They are assumed to be located near enough to the coast of the United States so that Congress may regulate them without significant international complications. They are also assumed to transmit electricity to shore by means of submarine cables.

The construction, operation, and maintenance of these OTEC plants will have various kinds of environmental impacts. Construction of OTEC plants could result in offshore pollution from vessels, from construction materials, and from human wastes. Operation and maintenance of plants also could cause pollution due to human wastes. All this pollution would be similar to that from other kinds of offshore facilities, such as oil drilling platforms.

Operation and maintenance of plants might also cause types of pollution unique to ocean thermal energy conversion. These might include discharge of substances, such as acids, used in maintenance; discharge of ions from exposed metal surfaces; or discharge of working fluid. The working fluid might consist of ammonia, Propane, or a fluorocarbon compound. Some fluorocarbon compounds are believed by scientists to cause depletion of the stratospheric ozone layer;[1] evidence that ozone depletion is taking place is mounting. If evidence to the contrary is not forthcoming, the use of these compounds as working fluids may be forbidden due to the risk that they will be released inadvertently into the atmosphere.

OTEC plants would create hazards of collision with ships, due to either ship navigational errors or the setting adrift of plants by human error or storms. In case of a collision, pollutants might be discharged from the ship or the plant or from objects with which a drifting plant collides. There would be some pollution from boats plying to and from OTEC plants. There would be adverse marine impacts from the laying of cables to bring the electricity ashore.

Construction, operation, and maintenance of OTEC plants would have numerous environmental impacts in the coastal zone. These include physical disruption of land, and air and water pollution due to construction and maintenance facilities. These facilities might include construction yards, boat harbors, and dry dock type of maintenance installations. There would be additional environmental impacts because of the presence of workers employed at these facilities. These employees will require housing, sewage disposal, schools, recreation, and other services with environmental consequences. There would be coastal environmental impacts from cables carrying electricity from OTEC plants and from transformers and transmission lines. If OTEC plants added greatly to the electric power supply of a state—such as Hawaii—there could be more substantial environmental impacts due to energy-stimulated economic growth. Again, these impacts would be similar to those of other offshore activities, such as oil development.[2]

Environmental Regulation

Environmental regulation will be necessary to minimize the harmful effects of the environmental impacts just described. Some of these regulations already exist. Others doubtless would be enacted to deal with special problems posed by OTEC facilities. The type of regulation will vary depending upon the state of progress in OTEC development.

The Research and Development Stage

Assuming the research and development program intended to produce a commercial OTEC technology is at least in part federally funded, as appears likely, the National Environmental Policy Act (NEPA) at a relatively early point will require the preparation of a programmatic environmental-impact statement. Section 102(2)(C) of NEPA requires that federal agencies prepare statements which discuss in detail the environmental impacts of and alternatives to "every proposal or recommendation for legislation or other major federal actions significantly affecting the quality of the human environment." These statements must be circulated to other agencies and the public for comment and considered in agency decisionmaking.

The impact statement requirements of NEPA do not require the impossible; all that is required is a detailed, good-faith assessment of a technology's likely environmental impacts and the alternatives. This is the kind of analysis a competent and prudent agency should perform as part of its own decision-making processes, so the additional costs of complying with NEPA should be small. In the case of the OTEC technology, the result of detailed, public assessment of environmental impacts may well be beneficial because OTEC facilities probably will have less severe environmental impacts than "competing" energy sources.

The leading case is *Scientists' Institute for Public Information (SIPI) v. Atomic Energy Commission,*[3] in which the United States Court of Appeals for the District of Columbia Circuit held that a NEPA impact statement had to be prepared on the Atomic Energy Commission's liquid-metal fast-breeder reactor program, a research and development program intended to prove the commercial feasibility of the liquid-metal fast-breeder reactor by 1980. The Council on Environmental Quality's NEPA *Guidelines*, promulgated after the *SIPI* decision, require agencies "engaging in major technology research and development programs" to "develop procedures . . . to determine when a program statement is required. . . ."

A federal program to develop a commercial OTEC technology would cost hundreds of millions of dollars and would entail environmental impacts which must be considered "significant" within the meaning of NEPA. The language of NEPA, the *SIPI* decision, and the CEQ *Guidelines* thus would require preparation of a programmatic statement at some point.

With respect to the key question of the timing of the impact statement, the AEC argued in the *SIPI* case that it would be premature to require immediate preparation of a comprehensive impact statement on the entire research and development program. The court declared, however, that:

> To wait until a technology attains the stage of complete commercial feasibility before considering the possible adverse environmental effects attendant upon ultimate application of the technology will undoubtedly frustrate meaningful consideration and balancing of environmental costs against economic and other benefits.[4]

The court applied a four-part test:

> How likely is a technology to prove commercially feasible, and how soon will that occur? To what extent is meaningful information presently available on the effects of application of the technology and of alternatives and their effects? To what extent are irretrievable commitments being made and options precluded as the development program progresses? How severe will be the environmental effects if the technology does prove commercially feasible?[5]

The agency responsible for preparing a programmatic impact statement on the OTEC research and development program will be the Energy Research and Development Administration, which is subsidizing the program under the Federal Nonnuclear Energy Research and Development Act of 1974.[6] Applying the four-part *SIPI* test, it seems apparent that the programmatic impact statement must be prepared *before* the decision to construct the first demonstration plant. Ideally, it should be prepared as soon as a respectable amount is known about the candidate OTEC technologies and their environmental impacts.

The programmatic impact statement should analyze the environmental impact of the OTEC technologies under consideration, including the impacts of the largest commercial OTEC industry which is foreseeable. It should identify areas in which knowledge about environmental impacts is lacking, so that research can be initiated before making an irrevocable commitment to a particular technology. The statement must compare with environmental impacts of the OTEC technologies with those of alternative energy sources in which research and development funds might be invested, such as solar energy, wind energy, thermonuclear fusion, and the breeder reactor, and weigh their benefits and costs.[7]

During the stage of OTEC research and development, ERDA will also be sponsoring research into other methods of energy production. Relative environmental impacts should play an important part in the allocation of R&D resources among competing energy sources, and formal assessments of environmental impact are required by NEPA to be prepared and considered in this decisionmaking process. It is therefore important for the federal government to conduct studies and collect information about those impacts as early as possible.

The Demonstration Stage

Following the R&D stage, and taking into account the research and development programmatic statement, the federal government may give serious consideration to subsidizing or permitting construction of one or more OTEC demonstration facilities. As part of this decisionmaking process, the programmatic impact statement should be updated unless only a short time has passed since the issuance or most recent updating of the research and development statement.[8] The demonstration-stage programmatic statement should discuss, in light of the most current information, the environmental impacts of the proposed demonstration technologies at both the demonstration stage and the commercial stage which may follow. Again, the environmental impacts of the OTEC technologies should be compared carefully with those of alternative sources of energy.

A separate section of this programmatic impact statement, or a separate statement or statements, should discuss in detail the environmental impacts of each proposed demonstration plant. The environmental impacts of alternative sites for the plant, including the impacts of associated onshore facilities, should be assessed in full.

At this point, federal and state environmental regulations other than NEPA will become applicable. Chapters 8 and 9 have set forth persuasive reasons why Congress will be required to enact new legislation regulating OTEC facilities within the jurisdiction of the United States. If Congress follows the pattern of the Deepwater Ports Act of 1974,[9] this statute and the regulations issued under it will regulate in detail the environmental as well as other aspects of offshore construction, operation, and maintenance of OTEC plants. It will probably prescribe a method for subsequent compliance with NEPA impact-statement procedures, at least with respect to individual plants.[10] It will probably establish a siting procedure for OTEC facilities which takes into account environmental impacts both offshore and onshore,[11] the interests of adjacent states,[12] and alternative ocean-surface and seabed uses.[13] It will probably prohibit polluting discharges or at least require use of the best available technology to minimize them,[14] and provide a method of compensation for environmental harm.[15] The statute should regulate the bringing of cables ashore from OTEC generating plants.[16]

The new federal statute might extend the reach of some state laws to OTEC facilities,[17] as do the OCS Lands Act[18] and the Deepwater Ports Act;[19] but if those statutes are any guide, the states will not play a significant role in environmental regulation offshore. However, onshore environmental impacts of OTEC facilities may be among the most significant.[20] During the demonstration stage, these will include construction and maintenance facilities and cable installations in the coastal zone, with accompanying pollution and urbanization. By the time specific OTEC demonstration plants are concretely proposed, all adjacent coastal states are likely to have adopted coastal zone management plans with federal assistance pursuant to the Coastal Zone Management Act (CZMA).[21] If the Deepwater Ports Act is indicative, Congress may require this as a condition of federal licensing for OTEC facilities.[22]

The CZMA provides that the coastal management plan must include a process for considering the national interest in the siting of energy-related facilities.[23] Federal permits for offshore facilities must be certified to be consistent with the coastal plan, thus giving adjacent states a veto over proposed OTEC sites, subject to an override by the Secretary of Commerce "in the interest of national security."[24] The Deepwater Ports Act flatly provides: "The Secretary (of Transportation) shall not issue a license without the approval of the Governor of each adjacent coastal state."[25] Congress might give the states an equally sweeping veto power over OTEC facilities, or might give federal agencies the same kind of overriding power as in the CZMA.

The identity of the "lead agency" responsible for preparing the programmatic environmental-impact statement at the demonstration stage will probably be determined by Congress in the new statute. The Department of Transportation, the Corps of Engineers, and the Interior Department are likely candidates. The Department of Transportation was given the lead with respect to deepwater ports.[26]

Commercial Stage

Assuming the demonstration plants operate successfully, there will follow a decision whether to permit commercial operation of OTEC plants. Because of the virtual certainty of comprehensive federal regulation and the significance of OTEC plants for the nation's energy needs, this decision doubtless will be made at the federal level. An updated programmatic environmental-impact statement should be prepared at the time of this decision. Based on the latest information, it would assess the environmental impacts of a mature commercial OTEC industry and compare them with those of alternative energy sources.

In addition, each individual commercial OTEC facility should be the subject of a site-specific environmental-impact statement which assesses in detail the environmental impacts of that facility and the alternative offshore and onshore uses. The preparation of these impact statements should be dovetailed with the permit procedures established by the comprehensive federal statute referred to above; it is likely that the statute will specify the point in the process at which the impact statement must be prepared.[27]

It is possible that at the commercial stage Congress will require by statute, or the responsible federal agency will prepare pursuant to NEPA, regional environmental-impact assessments, in addition to those on the whole program and on individual plants.[28] These regional statements would analyze alternative sites for OTEC plants on a regional basis, so that the full range of potential ocean uses could be taken into account and accommodated. Whether regional statements are desirable will depend in part on the number of OTEC facilities proposed off each U.S. coast.

As in the case of demonstration plants, the onshore aspects of commercial OTEC facilities—including construction, repair, and maintenance—will be subject to state regulation. It may be that, as is now proposed in the case of offshore oil drilling, the federal government will award special subsidies to states adjacent to OTEC facilities to enable them to do the necessary planning and otherwise deal with coastal impacts.

At the commercial-plant stage, and to some extent at the other stages, a host of federal and state environmental protection statutes other than those mentioned above will apply to OTEC facilities. Most of these are noted in Chapter 9. Applicable federal regulations include Corps of Engineers authority over obstructions to navigation pursuant to the Rivers and Harbors Act of 1899[29] and perhaps the Outer Continental Shelf Lands Act;[30] and Corps authority over dredge-and-fill discharges under the Federal Water Pollution Control Act of 1972 (FWPCA);[31] Environmental Protection Agency authority over water pollution under the FWPCA,[32] air pollution under the Clean Air Amendments of 1970,[33] and ocean dumping under the Ocean Dumping Act;[34] Coast Guard authority over design features,[35] and perhaps navigational and safety equipment;[36] Federal Power Commission authority over electric power transmission;[37] Department of Defense,[38] Department of Commerce,[39] and Department of the Interior[40] authority over potential competing uses of the oceans.

A variety of state regulations also would apply. They include coastal zone management statutes, perhaps modeled on the pioneering legislation now effective in California;[41] and "little NEPA's" like those on the books in California[42] and Hawaii,[43] which require state agencies to prepare environmental-impact statements. More direct state regulation offshore may also be possible, so long as it does not conflict with the federal regulatory scheme.[44]

These regulations generally are well described in Chapter 9 by James Higgins. In all probability, the procedure for coordinating compliance with them will be prescribed by the comprehensive federal statute generally regulating OTEC facilities, as occurred in the Deepwater Ports Act.[45]

Notes

1. *See, e.g.,* "Fluorocarbons and the Environment: Report of the Federal Task Force on Inadvertent Modification of the Stratosphere" (June 1975).

2. *See* P. Baldwin and M. Baldwin, *Onshore Planning for Offshore Oil: Lessons from Scotland* (1975).

3. 481 F.2d 1079 (D.C. Cir. 1973).

4. Id., at 1089.

5. Id., at 1094.

6. 42 U.S.C.A. §§ 4901 *et seq.* (Supp. 1975).

7. *See* Calvert Cliffs' Coord. Committee v. Atomic Energy Comm'n, 44 F.2d 1109 (D.C. Cir. 1971); Natural Resources Defense Council v. Morton, 458 F.2d 827 (D.C. Cir. 1972).

8. A NEPA impact statement must be prepared at each critical stage of the decisionmaking process. *See, e.g.,* Calvert Cliffs' Coord. Committee v. Atomic Energy Comm'n, *supra* note 7.

9. 33 U.S.C. §§ 1500 *et seq.* (Supp. 1975).

10. *See* Deepwater Ports Act of 1974, § 5(f); 33 U.S.C. § 1504(f) (Supp. 1975).

11. *See* id., § 6; 33 U.S.C. § 1505 (Supp. 1975).

12. *See* id., § 9(b)(1); 33 U.S.C. § 1508(b)(1) (Supp. 1975).

13. *See* id., § 6(a)(3); 33 U.S.C. § 1505(a)(3) (Supp. 1975).

14. *See* id., § 4(c)(5); 33 U.S.C. § 1503(c)(5) (Supp. 1975).

15. *See* id., § 18; 33 U.S.C. § 1517 (Supp. 1975).

16. The Deepwater Ports Act defines *deepwater port* to include pipelines. Id. § 3(10); 33 U.S.C. § 1502(10) (Supp. 1975).

17. *See* Chapters 8 and 9.

18. OCS Lands Act § 4; 43 U.S.C. § 1333(1973).

19. Deepwater Ports Act § 19(b); 33 U.S.C. § 1518(b) (Supp. 1975).

20. *See* Baldwin and Baldwin, *supra* note 2.

21. 33 U.S.C. §§ 1451 *et seq.* (1973).

22. That Act requires that the adjacent coastal state be making reasonable progress toward an approved coastal management program under the CZMA before a license may issue. § 9(c); 33 U.S.C. § 1508(c) (Supp. 1975).

23. 33 U.S.C. § 1455(c)(8) (1973).

24. 33 U.S.C. § 1455(c)(3) (1973).

25. § 9(b); 33 U.S.C. § 1508(b) (Supp. 1975).

26. Deepwater Ports Act § 5(f); 33 U.S.C. § 1504(f) (Supp. 1975).

27. Id.

28. *Cf.* Sierra Club v. Morton (D.C. Cir. 1975), suggesting the need for a regional impact statement on Northern Great Plains Coal Development.

29. § 10; 33 U.S.C. § 402 (1973).

30. § 4(f); 43 U.S.C. § 1333(f) (1973).

31. § 404; 33 U.S.C. § 1344 (1973).

32. 33 U.S.C. §§ 1251 *et seq*. (1973).

33. 42 U.S.C. §§ 1857 *et seq*. (1973).

34. 33 U.S.C. §§ 1401 *et seq* (1973).

35. 46 U.S.C. § 395 (1973).

36. *See* 33 C.F.R. §§ 144 *et seq.* (1974).

37. Federal Power Act, 16 U.S.C. §§ 791 *et seq*. (1973).

38. *See* 16 U.S.C. § 824(a),(e) (1973); 15 U.S.C. § 717(b) (1973); 18 C.F.R. Parts 32, 153 (1974); E.O. 10485 (Sept. 3, 1953).

39. *See* Marine Sanctuaries Act, 16 U.S.C. §§ 1431 *et seq*. (1973).

40. *See generally* OCS Lands Act, 43 U.S.C. §§ 1331 *et seq*. (1973).

41. Cal. Code Ann. §§ 27000 *et seq*. (Supp. 1974).

42. Id., §§ 21000 *et seq.*

43. Hawaii Rev. Stats. §§ 343-1 *et seq*. (Supp. 1974).

44. *See* Askew v. American Waterways Operators, 411 U.S. 325 (1973).

45. *See* Deepwater Ports Act § 4(c); 33 U.S.C. § 1503(c) (Supp. 1975).

11 Legal Aspects of Financing Ocean Thermal Energy Plants

John H. Riggs, Jr.

Introduction

One of the crucial aspects of the future successful development of commercial OTEC plants or devices, after the production of a tested demonstration model now proposed to occur by 1988, is the availability of financing. In light of the very substantial capital cost of OTEC plants, flexible means of financing their construction and use will be required in order to ensure their being built in sufficient numbers to justify the high cost of engineering and development investment by manufacturers and by the federal Energy Research and Development Agency, which is expected to finance the prototype and demonstration models of OTEC plants.

It has been estimated that the next 10 years will be very difficult times for existing capital markets in the United States and abroad. One respected source has stated that the United States will need the incredible sum of $4.5 trillion in new capital funds in the next 10 years.[1] That same source notes that such capital needs will be almost triple the new capital funds that have been raised over the past 10 years, often straining existing markets and methods severely and with universally recognized ill effects upon inflation in the Western world.

This chapter will attempt to suggest how currently available financing techniques could be applied to finance the construction, purchase, and use of OTEC devices after the development of a tested demonstration model and sufficient clarification of the legal status of OTEC plants in order to justify energy producers to commit to the construction and use of commercial OTEC plants. Such techniques are presently used by oil companies, shipping firms, and utilities to finance oil drilling and production platforms (particularly in the Gulf of Mexico and, more recently, in the North Sea), very large crude carriers (VLCC's), liquefied natural gas (LNG) carriers, natural gas and oil pipelines, ore carriers, and nuclear and other power plants, to name the principal kinds of equipment financed in the energy production fields. Many of these pieces of equipment are extremely costly, thus justifying the time and expense involved in the use of very sophisticated and complex financing techniques.[2]

It is assumed for purposes of this analysis that the operators of OTEC devices will be either U.S. privately owned public utility companies, other energy-producing companies, or chemical companies interested in the use of OTEC devices for the manufacture of certain chemicals (e.g., ammonia) or federal or state governmental agencies or specially created entities; that such devices will be used either in international waters beyond the present legal boundary of the United States (although possibly within an "economic boundary" of the United States which may be extended beyond the present 3-nmi limit, as was the case with respect to the continental shelf among the border states of the North Sea); and that they will be maintained and overhauled in U.S. ports. Consideration is given, however, to federal or state governmental ownership of the devices to the extent such ownership might influence financing techniques.

Discussion

The Financing of Privately Owned OTEC Plants

Unsecured Debt or Equity Financing by the Corporate Owner. Of the various financing methods available to a private corporate owner or user of an OTEC device, the most costly and generally the least complicated is the use of unsecured debentures or preferred or common stock of such owner or user, with the edge going to debt securities because of the tax deductibility of the interest thereon. In some instances, however, common stock also may become attractive to investors, when attractive dividends and relative financial health of the issuer join with the need to redress the debt/equity ratio required by financial markets. Because of the expected high cost of OTEC devices, it would probably be necessary for the corporate issuer or guarantor of such debentures or stock to have a high credit rating (at least double-A ratings by *Moody*'s and *Standard & Poor*'s may be necessary in order that the funds be available at an acceptable interest or preferred dividend rate, if the availability of funds for such purposes from private sources is limited by the projected competition for funds, as expected). While it is difficult to predict at this time the prevailing conditions in the financial markets when OTEC devices will need to be financed, it is safe to surmise that because of the high cost of OTEC devices and the competition for funds of such magnitude, their owners or users must be able to provide high credit standing to be able to attract sufficient funds from the public or institutional financial markets. Because of expected substantial other capital needs of the kinds of entities expected to own or operate OTEC devices, it will probably be necessary to provide that such debentures or preferred stock mature or otherwise remain unredeemed for long periods (20 to 30 years). Short-term borrowings from banks are unlikely to be involved, except as construction loans or as so-called bridge loans while permanent financing arrangements are being negotiated and completed.

If the owner or user is a constituent of a public utility holding company and the OTEC device is being used to create and transmit electricity to the United States, the terms of the debentures or stock offerings would have to be approved by the Securities and Exchange Commission (the SEC) under Sections 6, 7, and 12 of the federal Public Utility Holding Company Act of 1935, as amended,[3] and the rules of the SEC thereunder, and perhaps also by state authorities exercising regulatory jurisdiction over the company in question. The jurisdiction of the SEC over corporate members of a public utility group extends as well to secured bonds and other forms of financing, whether secured by tangible property or by other assets. Under Rule 50 of the SEC under such Act, public competitive bidding for the securities would generally be required, but there has recently developed a tendency toward negotiated purchases of such securities. It is customary for the SEC to impose restrictions on the terms of such securities in order to protect the public served by such utility group and the investors who purchase the securities. Furthermore, if the securities are placed publicly, usually through an underwritten public offering, they would have to be registered with the SEC under the federal Securities Act of 1933, as amended,[4] and the registration statement would have to be acceptable to the SEC.

When the owner or the user is unable to raise the necessary capital financing to construct or purchase OTEC devices because of its size or credit standing, because of contractual limitations on its borrowing powers, or because it wishes to avoid additional direct indebtedness competing with substantial financing needs for other capital assets to be purchased in the future, it could perhaps finance the OTEC devices by a secured arrangement by which the corporation would enter into long-range "take-or-pay" agreements under which the corporations or public utilities purchasing the electricity generated by the device, or the ammonia, or other products manufactured by the device, would agree to purchase an agreed amount of the electricity, or other product, at an agreed price for an agreed time, and would further guarantee that payment therefor would be made by it regardless of whether the full amount of such electricity or product is used or even received by it. The owner of the device would assign such "take-or-pay" contracts to a corporate trustee to secure the debentures or bonds issued by the device's owner without recourse against the owner (or even with recourse, as additional security, despite the issuer's inadequate credit standing alone). Pipelines, refineries, and electrical cooperative utility equipment are often financed in this manner. Thus, the credit of the purchaser of the electricity or other product is lent to the financing. Of course, the interest cost on debt financing of this sort is generally higher than in the case of a straight-debt financing of a corporate entity with high credit standing, because of the complications and doubts inherent in the "take-or-pay" agreements on the strength of which the indebtedness is issued and sold.

If OTEC devices prove themselves as dependable self-sustaining producers of energy with relative freedom from political and legal problems, replacing, at least to some extent, current fossil fuel energy producing systems, potential financiers will be attracted to investments in such devices instead of competing investments. Furthermore, such a proven success could lead to development of project financings under which more faith is placed by lenders in the earning capability of the project being financed (if the investor or financier has a superior right to earnings) than in the user of the device.

Ship Mortgage Financing. OTEC devices are expected to spend most of their time on the high seas, outside the jurisdiction of the United States or any other country, and accordingly no national mortgage law will be available in such cases to assure a perfected mortgage on behalf of prospective lenders or investors.[5] Since the OTEC devices will be either moored to the seabed or will float on the surface under self-propelled power, they may, however, qualify for treatment as vessels for purposes of the federal Ship Mortgage Act of 1920, as amended,[6] or the ship mortgaging laws of so-called flags-of-convenience jurisdictions, such as Liberia, Panama, Costa Rica, and Honduras.

To qualify for a ship mortgage under U.S. law, the OTEC device in question must constitute a "vessel of the United States," i.e., a water craft or other "artificial contrivance" which is "used or capable of being or intended to be used as a means of transportation on water"[7] and which is documented under U.S. law.[8] To be documented under U.S. law, a vessel must be (i) either built in the United States, or (A) built abroad and certified by the U.S. Coast Guard as "safe," and (B) used only for trade with foreign countries, and (ii) wholly owned by U.S. citizens or nationals.[9] In the past, stationary drilling rigs and drilling ships built in the United States and owned by Americans have qualified as "vessels of the United States" on which U.S. ship mortgages could be obtained for financing purposes. Nevertheless, the question remains whether the various proposed designs of OTEC devices will qualify for treatment as "vessels" for purposes of ship mortgage financing under U.S. law (see Chapter 8) and whether application of such law to OTEC devices for such purposes would survive an attack by unsecured creditors in a case where bankruptcy or reorganization occurs. There is little question that clarification of U.S. law in this respect would assist in assuring the availability of secured financing techniques to OTEC development.

The statutory preferred mortgage created by the Ship Mortgage Act of 1920, as amended, gives the holder or holders of the mortgage a priority in the distribution of the proceeds from the sale of the vessel over all other claims against the vessel except preferred maritime liens and certain expenses and costs allowed by the court.[10]

With respect to OTEC devices as to which a "flag of convenience" may be deemed preferable to U.S. documentation,[11] the national law of the country of documentation would generally govern whether devices of this nature would qualify for treatment as "vessels" for purposes of ship mortgaging under such laws. U.S. courts are required to recognize the priority established by a duly perfected preferred ship mortgage created under foreign laws.[12]

In summary, if the OTEC device in question is determined to fall within the definition of "vessel of the United States" referred to above or "vessel" under foreign ship mortgage laws, the ship mortgage would be available in many instances as a security interest to enable financing of OTEC devices, notwithstanding such devices being located outside the jurisdiction of any country or in an area where no other recognized security interest or mortgage law would apply. The value of such security in facilitating such financing depends largely on the establishment of a resale market for such devices upon a default in the loan agreement or other credit instrument under which the device is financed, and until the ability to assure a substantial resale value is assured, the availability of such a security interest is likely to be only of marginal utility in financing. Furthermore, recourse to a mortgage will limit financings relying on the mortgage to only a portion of the cost of the OTEC devices, since the lenders will want to be sure of an adequate resale value on default and will accordingly probably limit such financings to around half the construction cost of the device.

In addition to ordinary private ship mortgage financing, OTEC devices may be eligible for U.S. federal government-assisted ship mortgage bond financing under Title XI of the Merchant Marine Act of 1936, as amended.[13] Under Title XI, the federal government will currently guarantee ship mortgage bonds (with the government as the mortgagee) issued in connection with the financing of new vessels (interpreted to include oil drilling rigs of the semisubmersible and "jack-up" types) of U.S. registry and ownership[14] which are also manufactured in the United States. The purpose of Title XI, stated very generally, is to promote and support U.S. shipbuilders and the creation of a U.S. merchant marine fleet by lowering the financing costs of U.S.-built vessels through the grant of federal government full faith and credit guarantees of the debt obligations purchased by public or institutional investors. To obtain such a guarantee, the borrower must submit an application to the U.S. Maritime Administration (MarAd) of the Department of Commerce demonstrating, among other things, that the borrower has the operating ability and experience necessary to operate the "vessel" properly and that it has the financial resources necessary for such operation. To meet the financial resources requirement, the borrower need not have a particularly high credit standing; such test is met on individual criteria on a case-by-case basis. If MarAd grants the application, the guarantee is limited by law to a statutory 75 percent or 87.5 percent of the "actual cost" of the vessel, as defined in Title XI. If the application is granted, the federal government guarantee substantially lowers the interest cost of the financing, although the government receives a statutory fee of 1/4 to 1/2 percent of the principal amount of the ship mortgage bonds which it guarantees and despite the considerable complexity and documentary burden which dealing with the federal government in this context imposes.

Whether Title XI would apply to OTEC devices is not clear and depends in large part upon presently unpredictable political considerations. As a recent

MarAd general counsel has pointed out, the stated policy of the Merchant Marine Act of 1936, as amended, is to limit guarantees to vessels which are modern additions to the merchant marine capable of serving as naval auxiliaries in time of war.[15] Whether OTEC devices would meet these and other governmental tests is anyone's guess. Depending upon the status of the world shipping business when application for extension of Title XI is first made in connection with OTEC devices (the market is presently in a severe depression with the result that American shipbuilders are searching for new contracts), U.S. shipping firms competing for the limited Title XI guarantees authorized by Congress may make extension of Title XI to OTEC devices politically difficult. In any case, the Congress will probably have to expressly authorize extension of Title XI to OTEC devices, which were not originally contemplated by the Merchant Marine Act of 1936 or amendments heretofore made to it, in order to assure ready availability of such government guarantees to facilitate financing.

Both the Ship Mortgage Act of 1920 and Title XI have the considerable advantage of providing established legal systems for secured financings of equipment broadly comparable to OTEC devices and, in the case of Title XI, a government administration already experienced in administering the federal guarantee program. Of course, political considerations may militate in favor of the creation of a new federal agency to administer federal financial assistance for the construction, purchase, and use of OTEC devices or the assignment of such task to ERDA or some other existing agency.[16]

Before leaving the subject of ship mortgages and Title XI federal government guarantees, it is worth noting that another factor influencing the financing of OTEC devices, especially where private placements of securities are concerned, is the freedom of regulated U.S. banks and insurance companies (the institutional investors which have thus far been instrumental in the financing of the Trans-Alaska oil pipeline and other recent major energy-related equipment) to purchase the securities in question. For purposes of this chapter, it is sufficient to state that banks, whether national or state, and insurance companies are subject to complex regulatory limitations on foreign investments and unsecured investments.[17] Financing of OTEC devices should be framed in such a way as to meet the broadest categories of such regulatory limitations in order to be available to the broadest spectrum of institutional investors. However, past experience in offshore oil drilling and production rig financings has shown that, with ingenuity, such regulatory tests can be met in a manner which will open the financing to as many potential institutional investors as possible. The availability of settled market techniques for financing OTEC devices would permit such devices and their users to compete equitably in the market place during the expected future credit squeeze.

Lease Financing. Because of recent high long-term interest rates and inflationary factors, many oil companies and utilities have in the past few years turned increasingly to lease financing of oil tankers, oil rigs, and utility equipment. The basic principle of lease financing is that U.S. tax advantages (accelerated depreciation and, generally, investment tax credit) otherwise available to the owner and user of the equipment (but sometimes not fully usable because of other available deductions or credits) are traded to U.S. financial institutions which can use such tax benefits, in return for a lower net financing cost for the equipment over most of its useful life.

The principal financing technique used in leasing is the so-called leveraged lease.[18] In a leveraged-lease financing, the equipment is delivered by the manufacturer to the lessor, usually a commercial bank acting as trustee for one or more financial institutions as trustors, who are the beneficial owners of the equipment. The trust is established in such a manner as to qualify as a partnership for federal income tax purposes, so that each trustor-owner is able to take its pro rata share of depreciation benefits, interest deductions on the nonrecourse secured debt issued by the trust to finance up to 80 percent of the cost of the equipment, and any investment tax credit. The trustee enters into a net, "hell-or-high-water" lease with the lessee (upon whose creditworthiness, or that of the lessee's parent, which guarantees the lessee's obligations, the transaction is based). Simultaneously, the trustee issues nonrecourse equipment trust notes or similar securities usually secured principally by an assignment of the lease rentals and a mortgage on the equipment. The lease rentals are sufficient to service principal and interest on such notes or other securities and to provide the trustors with some cash flow. All the tax benefits (except sometimes the investment tax credit, which, if available, may be retained by the lessee) flow to the trustors. In fact, the rate of return to the trustors is usually substantially greater than that which they would have in an ordinary financing with the same borrower.[19]

The principal advantages of lease financing to the borrower are: (1) the lower net cost of the financing; (2) such financings are often not shown on the balance sheet of the borrower; (3) lease financing may not be subject to limits imposed under other borrowing instruments of the borrower on straight or secured debt financing; (4) the tax deductibility of the lease rentals; and (5) the avoidance of massive capital expenditures in the year the equipment is acquired. In addition to the tax shelter obtained by the trustors, who own and lease the equipment through the trustee, the trustors can expect in many cases to receive an important nontax advantage in the form of the residual value of the equipment at the termination of the lease. But such residual value becomes very difficult to predict where certain oil rigs are concerned, because of moving costs, and the residual value (other than scrap) of OTEC devices is likely to be similarly unpredictable.

Since the tax advantages passed on to the trustors are key to the transaction, rulings are almost invariably obtained from the Internal Revenue Service before the transaction closes. The lease must qualify as a true lease and not possess the attributes of a conditional sale.[20] After long deliberations on the subject, the IRS recently published guidelines for issuance of its rulings in leveraged-lease financings.[21] Under such guidelines, advance rulings will be issued if (1) the trustors themselves invest a minimum of 20 percent of the cost of the equipment, (2) the fair market value of the equipment at the termination of the lease will be at least 20 percent of its original cost, (3) at the end of the lease term the remaining useful life of the equipment will be longer than one year or 20 percent of the originally estimated useful life of the property, (4) the lessee and its affiliates must not have a right to purchase the equipment at the end of the lease term for less than its then fair market value and must not make any investment in, or guarantee any equipment trust note issued in connection with, the equipment, and (5) the trustors must receive a profit from the transaction other than the tax benefits mentioned above. Certain other requirements must also be met, and the ruling procedure has now become quite cumbersome. In the case of the first major leveraged-lease financing of an oil drilling and production platform in the North Sea in 1975, the Internal Revenue Service refused to rule on the availability of the platform for leveraged-lease financing because of technical questions raised by the service on the residual value of the platform after the expiration of the lease (based on current technology), thereby probably eliminating for the time being the availability of such a financing technique for the costly equipment necessary to retrieve oil from the hostile environment of the North Sea.

With respect to leveraged-lease financing of OTEC devices, special problems arise because of the use of the devices outside the continental limits of the United States. First, to qualify for the investment tax credit, the investment must have been made in equipment to be used predominantly within the United States.[22] However, equipment used predominantly outside the United States nevertheless qualifies for the credit if it is a "vessel documented under the laws of the United States which is operated in the foreign or domestic commerce of the United States."[23] Subject to the construction and ownership requirements for U.S. registration of vessels, OTEC devices in many cases could fall within this exception so that the investment tax credit would apply. Second, since OTEC devices will be used outside the United States, lease rentals derived from such foreign use will constitute "foreign-source income" for U.S. federal income tax purposes.[24] Since the credit for taxes paid by the trustors to various foreign governments is limited to the same proportion of the U.S. income tax against which such foreign tax credit is taken that total foreign-source income bears to total income of the trustor from all sources (U.S. and foreign), any reduction in foreign-source income would generally result for most potential trustors in a corresponding reduction in the use of foreign tax credits to the extent such reduced foreign-source income would be insufficient to permit full credit for all foreign taxes paid.[25] Depreciation deductions with respect to the equipment must be applied against foreign-source income of the trustor from other sources in determining the availability of foreign tax credits from other sources. The depreciation deductions from the lease financing may offset so much of the trustor's other foreign-source income that otherwise available foreign tax credits are lost. Therefore, unless the trustor has no foreign-source income from other business conducted by it (e.g., interest on Eurodollar loans to foreign borrowers), its participation as a trustor in a leveraged-lease financing of an OTEC device could limit the use of foreign tax credits otherwise due from other foreign business. However, Section 861(e) of the Internal Revenue Code permits a U.S. trustor-owner to elect permanently to treat income (including gain or loss on disposition) from the lease to another (unaffiliated) U.S. person of a "vessel" built in the United States as U.S.-source income, even where the vessel is used outside the United States. Nevertheless, the foreign tax credit problem has been generally responsible for limiting the participation (as trustor-owners) of major U.S. multinational banks in leveraged-lease financings of oil tankers and drilling or production platforms to be used outside the United States, and the same problem is likely to somewhat limit, under current law, the use of leveraged-lease financing for OTEC devices.

The Financing of Government-Owned OTEC Plants

Another potential financing device for OTEC plants is the use of corporate or public entities owned by the federal government or state governments which accomplish the financing through revenue bonds issued by such corporation or entity and backed either by the "moral obligation" of the government which owns the issuer-owner or by the constitutional full faith and credit of the government in question (the latter would clearly result in lower financial costs, given recent reactions to the financial problems of The City of New York, New York State's Urban Development Corporation, and other municipal borrowers, although much depends also on which government is backing the bonds). Since the revenue bonds of such a public corporation or government entity would presumably be exempt from federal income taxes and the taxes of some states,[26] substantial interest costs on the financings could normally be expected to be saved. The Congress has in the past favored the development of local small industry, investment in pollution control equipment in factories, and the building of certain municipal facilities by permitting state and local authorities to issue so-called industrial-development or pollution-control revenue bonds, which are issued on a limited-recourse basis by a specially created state or local authority whichh owns the property, with the security (and on the credit), of a lease, conditional-sale agreement, or loan agreement with the user of the facilities in question, but qualifying for federal tax exemption.[27] With this precedent the Congress may also be willing to extend such favorable tax treatment to interest on state or local governmental authority-issued bonds used to finance OTEC plants.

To further lower the interest rate on such securities, the government entity could take, and assign for the benefit of its bondholders, "take-or-pay" contracts from utility or other users of the electricity or other product generated by the OTEC device to widen the credit basis for the bonds. Fully 100 percent of the cost of the OTEC device could be financed (as in lease financings).

An additional advantage of financings backed by the credit of the federal government or state or local entities is that the securities in question may be offered and sold to the public without incurring much of the additional cost of registration with the SEC.[28]

Because of the strain which use of government entities would place on other financing needs of such government in the "credit crunch" expected to develop in coming years, it may be politically difficult for this sort of financing method to be used widely, despite the cost savings and the potential profit for the general public which would develop as the OTEC devices produce energy. Government guarantees for privately owned OTEC devices and tax exempt status of interest costs in connection with the use of revenue bonds backed by the private user rather than the titular governmental owner would be less demanding ways of fostering OTEC devices without serious impact on the government's borrowing powers for other purposes.

Mixed Forms of Financings

OTEC devices will probably lend themselves easily to combined or hybrid forms of financings. For example, the Lockheed design provides for a central core structure and four or more attached (and removable) power modules. Much as in the case of modern passenger aircraft financings (where the jet engines are often separately financed) and nuclear fuel financings, OTEC devices could be financed, in part or in whole, by several of the above forms of financing, thus broadening the opportunity to place the securities with different financial institutions or other financing sources with varied objectives (term of financing, tax advantages, etc.).

Furthermore, various forms of credit standing could be combined, much as in the case of the current vogue of so-called project financing in mainly less-developed countries, to obtain less expensive financing; for example, ship mortgage security coupled with the assignment of a long-term net lease (or charter) of the OTEC device to a user which would thereby lend its own high credit standing. As in project financings, the very technical and commercial success of the OTEC device could lend itself, once proven, to facilitating lower financing costs.

Conclusion

Of the different financing techniques discussed above, it is questionable and in some cases quite unlikely that ship mortgaging, Title XI federal government guaranteed ship mortgage financing, leveraged-lease financing, and tax-exempt industrial-revenue bond financing would be available for financing OTEC devices under present law. A statutory amendment is probably necessary to assure to prospective lenders the availability of the Ship Mortgage Act for secured financings of OTEC devices. Unless the Congress and the Executive or the interested states create or clarify the availability of preferred financing forms for OTEC devices, such as tax-exempt interest, federal government guarantees, and secured financings, it is likely that despite the substantial ecological and operational advantages which preliminary studies indicate exist with respect to OTEC devices as contrasted with competing energy sources, such devices will actually be at a disadvantage in the looming competition for capital funds, with ships and pollution-control devices having clear access to favored treatment. OTEC devices face very substantial technical and economic risks before their success as an alternate economically viable energy source can be assured. In addition, the legal climate in which OTEC devices will be operated is uncertain.[29] To reduce these risks, at least until OTEC devices prove themselves in the economic, technical, and legal spheres, the responsible government authorities will have to consider whether OTEC devices warrant priority treatment, either on a par with, or superior to, the kinds of financing currently favored by statutory public policy. And as other competing energy sources are developed, the question will arise as to whether indirect subsidies, through according preferential financial status to OTEC devices and equipment used in tapping such other sources, will be necessary to develop new technologies which will free us from unwholesome dependence on unstable foreign energy sources that will eventually be exhausted.

Notes

1. *Business Week*, September 22, 1975 (pp. 42-115).

2. For example, an approximately 250,000 dwt VLCC ordered in 1971-72 cost generally from \$35 million to \$40 million to build for delivery in 1974. The "Condeep" drilling and production platform called "Beryl A Platform" built in Norway and installed during the summer of 1975 in the Beryl field in the U.K. sector of the North Sea will have cost approximately \$230 million, and similar platforms under construction will cost considerably more. LNG tankers presently being built in the United States may cost more than \$115 million each.

3. 15 U.S.C. §§ 79-79z-6.

4. 15 U.S.C. §§ 77a-77aa.

5. Even if the United States were to extend its territorial limits beyond the present 3 nmi to an "economic zone," and OTEC plants were located in such zone (e.g., off Hawaii), there is currently no statutory scheme for the perfection of a security interest or mortgage in such equipment, other than federal ship

mortgaging provisions discussed below. Section 9-102 of the Uniform Commercial Code (UCC) provides that the UCC only applies to personal property "within the jurisdiction of this state," and Section 9-103(2) would not seem to apply to OTEC devices so as to otherwise make the UCC available. For the U.S. Supreme Court's decision on the federal government's preemption of state rights over the seabed and subsoil between the 3-nmi limit and the continental shelf, *see* United States v. Maine, 420 U.S. 515 (March 17, 1975).

6. 46 U.S.C. §§ 911 *et seq*.

7. 36 U.S.C. § 801.

8. 46 U.S.C. §§ 911(4), 921, 922.

9. 46 U.S.C. § 11. *See* McDonald, "Documentation and Transfer of Vessels; Transfer of United States Vessels to Aliens," 47 *Tulane L. Rev.* 511, 516 (1973).

10. *See generally* Smith, "Ship Mortgages," 47 *Tulane L. Rev.* 608 (1973). A detailed description of the formalities and limitations on ship mortgages under the Ship Mortgage Act of 1920 is outside the scope of this chapter, and reference should accordingly be made to the above comprehensive article and other source materials cited therein.

11. For a general discussion of various forms of ship financing, *see* Mahla, "Some Problems in Vessel Financing—A Lender's Lawyer's View," 47 *Tulane L. Rev.* 629 (1973).

12. 46 U.S.C. § 951.

13. 46 U.S.C. §§ 1271-81. *See generally* Cook, "Government Assistance in Financing—Title XI Federal Guarantees," 47 *Tulane L. Rev.* 653 (1973).

14. Within the meaning of Section 2 of the Shipping Act of 1916, as amended; 46 U.S.C. § 802.

15. Cook, *supra* note 13 at 669.

16. On October 10, 1975, President Ford submitted legislation to the Congress to create the Energy Independence Authority (EIA), a new governmental corporation which would provide loans, loan guarantees, and other financial assistance to private-sector energy Projects. *See* B.N.A. *Business Policies* (October 10, 1975), pp. A-13 to A-15, B-1 to B-5; U.S. Senate bill S. 2532 and House of Reps. bill H.R. 10267, 94th Congress, 1st Sess. EIA would have $25 billion in equity and $75 billion in debt.

17. *See, for example*, Sec. 81 of the New York Insurance Law, by which most major U.S. insurance companies test the legality of their investments.

18. On leveraged leasing generally, *see* Stiles and Walker, "Leveraged Lease Financing of Capital Equipment," 28 *Business Lawyer* 161 (1972); "The Powerful Logic of the Leasing Boom," *Fortune* (November 1973); Angermueller, "Miscellaneous Ship Financing," 47 *Tulane L. Rev.* 725, 726-31 (1973). Much has been written on this subject in the last two years by accountants, lawyers, and businessmen.

19. For a schematic presentation of a leveraged lease, *see* Chapter 8.

20. Rev. Rul. 55-540 (1955-2 Cum. Bull. 39) contains some guidance from the IRS on what constitutes a true lease.

21. Rev. Proc. 75-21 (T.I.R. 1362, April 11, 1975); Rev. Proc. 75-28 (T.I.R. 1371, May 5, 1975).

22. Internal Revenue Code (IRC) § 48(a)(2)(A).

23. IRC § 48(a)(2)(B)(iii).

24. IRC § 862(a)(4).

25. IRC §§ 901 and 904(a)(2); Stiles & Walker, *supra* note 18 at 168; Angermueller, *supra* note 18 at 729.

26. The exemption from federal income taxes for a state or municipal governmental entity's obligations is created by IRC, § 103(a)(1); Obligations of a federal government entity would not be exempt from federal income tax unless the Act of Congress establishing such entity so stipulated, IRC, § 103(b). Most states exempt from their state or local income taxes the debt obligations of public entities organized in such state.

27. IRC § 103(c).

28. 15 U.S.C. § 77(a)(2).

29. *See* Chapters 2 and 8.

Appendix: Reports of the Working Groups Workshop on Legal, Political, and Institutional Aspects of Ocean Thermal Energy Conversion

Report of Working Group I: International Legal Aspects of OTECs

J. Christian Kessler, Editor

The discussions of Working Group I on the International Political Aspects of OTECs with Panel Members H. Gary Knight, Ann Hollick, and Byron J. Washom were focused on two questions. First, how will OTECs conflict with other uses of ocean space? Second, what are the legal and political implications of plant design? These questions were considered in terms of the current international regime of the oceans and anticipated outcomes of the third session of the UN Conference on the Law of the Sea (LOS III). Obviously much of the discussion was highly speculative, given both the uncertain future of LOS III and the rapidly changing uses of the oceans.

The working group identified a number of areas of agreement. While there were some disagreements, these concerned aspects of the future which are so speculative that no one appeared willing to press a point. The most important areas of agreement concerned jurisdiction to deploy OTECs, the reaction of less developed countries (LDCs) to such a deployment, and possible conflicts with other users of ocean space.

Any attempt to put an OTEC at sea under the current regime would have to be justified by the reasonable-use doctrine, and thus might not be acceptable to all states. If the Single Negotiating Text (SNT) is accepted in something similar to its present form, then jurisdiction within the economic zone for economic purposes is clearly with the littoral state. On the high seas beyond national jurisdiction the situation is still ambiguous. Although this would not be a problem for the use of OTEC as a power source plugged into a country's power system, operation of OTECs on the high seas as power sources for sea-based activities would broaden the market. However, some do not consider this to be a problem until there is a significant number of OTECs, which would not be for at least another 25 years.

There was complete agreement within the working group that the reaction of LDCs to OTECs would be the key to their acceptance or rejection on the high seas. The working group isolated three factors which will drive the LDC reaction to OTECs. First, will the LDCs be consulted on the development and deployment of OTECs, or faced with a *fait accompli* by the United States and other developed states? Second, what sort of policy will the United States develop for technology transfer (will the LDCs just be permitted to buy the finished product or will they be permitted to obtain some of the technology involved)? Third, how will the power from OTECs be used? If it is used for supporting industries at sea which are too "dirty" for the developed states to permit at home, how will the LDCs be likely to react to these uses of OTECs? The strength of the

LDCs was agreed to be their power to raise the question of who owns the resource of thermal energy in the water column. While it currently appears to be *res nullius*, if the LDCs are dissatisfied with the developed countries' use of OTECs, they are likely to fight for inclusion of deep-sea thermal gradients in the common-heritage concept and for regulation of their use by an international authority.

Even if the basic problem of ownership of the resource does not develop, conflicts will develop between OTECs and other uses of ocean space or ocean water (for either cooling or waste disposal). These problems of conflicting use would, by general agreement, be most difficult close to shore (within the 200-mi economic zone or over the continental shelf). Navigation is likely to generate the most widespread, and tractable, problems. The discussion about conflicts with fishing were very speculative because of the uncertainty about location of OTECs, the effects of artificial upwelling, and the degree conflicts, such as interference of artificial upwelling, and the degree to which there would be artificial upwelling. Most other conflicts, such as interference with ocean mining, petroleum production, military operations areas, or other uses of ocean water for cooling or waste disposal, which are spatial in nature, could be resolved through the adoption of a leasing or licensing system.

The working group identified several lacunae in current United States policy, and a number of issues which require further research. These suggestions can be divided into those which relate to the management of OTEC development and those which relate to the management of OTEC use. It should be borne in mind, however, that a number of the research recommendations described below would more effectively be considered after an OTEC prototype is in use.

Policy Recommendations

Management of OTEC Development:

1. A funding commitment should be made for a given number of years and a "go/no go" decision made at the end of that time. This recognizes that the federal government's investment and commitment in OTEC will precede industry expenditure of its own resources.
2. ERDA should establish an explicit policy regarding technologies which have domestic application but which will probably be used principally for export. The use of the U.S. and foreign patent systems as a basis for the transfer of technology should be examined.

Management of OTEC Use:

1. LDCs should be kept informed of U.S. progress in OTEC development, and interested LDCs might be invited to participate.

2. A policy should be established regarding the use of technology development and technology transfer as tools of foreign policy. This should apply to transfers both to developed and developing countries.

Research Questions

Management of OTEC Development:

1. To what degree is the development of OTEC technology a public-policy problem?
2. Do market imperfections mitigate against the development and introduction of alternative energy sources, particularly those involving exotic fuels?
3. What are the legal implications of different design decisions? Knowing the implications of different designs might change the balance among the alternatives, or suggest different designs which are equivalent technically but quite different legally. Some suggested areas of research are:
   a. Aids to navigation
   b. Legal permissibility of mooring versus grazing
   c. Influence of technology on high seas versus economic zone deployment
   d. Use/misuse potential of effluents
4. What are the implications of technical decisions made during development for the transfer of the technology to other countries, especially LDCs?

Management of OTEC Use:

1. What is the potential for effective use of technology transfer as an instrument of foreign policy? How should such transfers be licensed?
2. The potential problems caused by conflicting uses of ocean space should be examined in detail, focusing on the following issues:
   a. How will power from OTECs be used?
   b. How have the United States and other states handled (or not handled) the problem of intensive multiple uses of ocean areas (e.g., Gulf of Mexico and North Sea)?
   c. What legal issues are involved in establishing special intensive-use areas on the oceans?
   d. How should rights be allocated for those resources which are currently or were formerly *res nullius*? (e.g., how should fishermen be indemnified for damage to their fishing grounds, if it occurs?)
   e. What are the problems in managing an economic resource zone in which some uses are subject to national control (all economic activity) while other uses are subject to international control (navigation)?

Report of Working Group II: Regulatory and Liability Aspects of OTEC

John McLain, Editor

The Working Group met with panel members J.D. Nyhart, C.G. Hallberg, and James C. Higgins. The general consensus reached by the Working Group was that OTEC, however defined, or wherever located, did not fall neatly into any existing, traditional legal framework. There are a range of other ocean devices which share this characteristic; for example, offshore nuclear power plants, aquaculture, data acquisition systems, and deep-sea mining devices.

One of the central questions raised in dealing with the regulation of OTEC is whether it will be considered a vessel or a nonvessel. As the panelists pointed out, the Group's original assumption was that OTEC would be considered a vessel and the papers were approached from that standpoint. On the other hand, more recent suggestions indicate that for certain uses it may not be considered a vessel. If it is considered a nonvessel, many aspects of construction, inspection, documentation, regulation, and liability will be affected. Of particular importance would be OTEC's applicability to maritime law; its eligibility under the Preferred Ship Mortgage Act and the Ship Owner's Limitation of Liability Act; its standing under the Outer Continental Shelf Lands Act and the Deep Water Ports Act; its regulation by the EPA; and its relationship to various laws governing customs, the environment, and fishing within the contiguous zone and beyond.

The economic activity engaged in by an OTEC device, from the production of electricity to refining and manufacturing, will have an impact on its definition and standing under maritime, international, and national law. If an OTEC device is involved in explorative and/or exploitative activities concerning the seabed, this will affect its treatment under law. The Outer Continental Shelf Lands Act and the Deep Water Ports Act both extend federal and state law over the subsoil, seabed, and port facilities beyond the territorial sea. Federal and state law will also apply differently to various types of economic activities conducted on or by an OTEC device. Presently, state law is used to fill in the gaps in existing national legislation. The applicability of state law becomes more problematic as OTEC is placed further from the coast and its ties to a state become more tenuous. It was the view of the Working Group that the regulatory and liability aspects tend to be intertwined and that there is a vital need for guiding regulations if industry is to continue to be involved in OTEC's development.

As a means of simplifying the federal application procedure for dealing with OTEC, the Working Group recommended that a lead agency be created. This lead agency is envisioned as a one-stop process for the approval of the construction and operation of OTEC. As such, it would have the responsibility for coordinating the efforts of the various government agencies which would be involved in any approval process.

Three suggested means of modifying the approval process were: (1) limit the number of steps necessary for approval; (2) restrict the number of opportunities whereby interested parties may intervene in the approval process without removing the qualitative ability to intervene; and (3) create statutory limitations which must be met by the various government agencies when acting on the application.

One member of the Working Group suggested the need for new legislation. This legislation would: (1) contain a declaration establishing that an OTEC platform is to be considered a vessel; (2) establish that the environmental effects for the first generation of OTEC platforms not only meet existing U.S. requirements concerning discharge temperatures and other potential environmental hazards, but that baseline data studies are required; and (3) recommend a new IMCO requirement for the discharge of antifouling chemicals into the sea, and a uniform system of navigation aids and markings to be installed on all OTEC platforms.

A variety of opinions were expressed within the Working Group concerning the feasibility and urgency of OTEC and legislation pertaining to it. As the Working Group progressed, a general consensus developed dealing with the need to ensure that the legal solutions and the scientific/technical aspects of OTEC advanced in a parallel fashion.

One alternative mentioned was for the passage of legislation to stimulate private investment in OTEC. A number of arrangements were suggested, including federal loan guarantees and a federal subsidy program.

Report of Working Group III: The Domestic and International Environmental Aspects of OTEC

Cathryn M. Dickert, Editor

Panelists Robert E. Stein and Thomas B. Stoel led the discussion in Working Group III concerning the international and domestic environmental aspects of OTEC. The objective of this Working Group was to consider recommendations concerning the environmental procedures that the Energy Research and Development Administration (ERDA) could follow in the near future, as well as further along in development of OTEC. Additionally, the Group sought to outline future research priorities of an environmental nature which should be undertaken as part of the development of OTEC.

One of the first recommendations urged ERDA to begin the process of developing the data for and then formulating a programmatic environmental impact statement. This was felt necessary due to the fact that at present there are a number of questions—some based on speculation concerning environmental hazards surrounding OTEC technology. In proceeding with the programmatic statement, which would consider the effects of the development of the entire OTEC program at an early stage, the Council on Environmental Quality (CEQ) *Guidelines* were offered as a model which would help to limit and determine when a programmatic statement is necessary. It was the general consensus that, at this time, it would be helpful to ascertain what considerations should be included in this statement, as well as provide an outline limiting the scope of the initial investigation. The group pointed out several important factors which should be considered in the initial statement. They included a definition of OTEC technology, the identification of the primary uses of the OTEC facility, and an enumeration of the environmental hazards involved. Favorable comments were made concerning the use that the programmatic environmental impact statement at this time would have since (1) it would provide a mechanism whereby all current information concerning OTEC would be identified, collected, and made available; (2) it would provide a source of information to appropriate domestic agencies of the federal government concerned to the nongovernmental sector, as well as provide an opportunity to foreign governments interested in the technology to review the effect the development of OTEC will have on their interests; (3) it would serve to identify where the gaps exist in the data on the environmental impacts of the technology and provide an idea where additional research needs exist; (4) it would also serve as a useful tool in the comparison of OTEC with other alternative sources of energy production.

The Group next discussed limitations of the scope of a programmatic statement. It was generally agreed that the first step necessary would be to define OTEC and identify its primary uses. For example, the utilization of OTEC as a potential source of energy for offshore industrial activities, such as bauxite refining, was discussed. The Group felt that such uses were beyond the scope of immediate concern of the programmatic statement and that such "secondary" uses would be the focus of demonstration or commercial impact statements at a later time. Factors, they decided, to be considered presently should relate to the OTEC facility itself, such as decisions concerning an environmentally sound technique for dealing with biofouling or development of a safe working fluid. Additionally, the effects an OTEC facility, or a large number of them, would have on the atmosphere and the ocean environment, including fisheries, should be considered as well as conflicts involving effects on existing navigational routes. There was also considerable apprehension voiced concerning possible climatic or weather modifications which might result from changes in the temperature of the Gulf Stream.

The enumeration of such hazards would require long-term studies. However, such questions and considerations led to agreement that a programmatic environmental impact statement would investigate such matters and serve as an effective means of dispelling much of the speculative information afloat concerning OTEC. It could also provide ERDA with necessary information to formulate a decision whether to continue support through additional research in OTEC as a viable alternative source of energy.

Next, the discussion considered the international implications that OTEC plants might have. Since the United Nations Conference on the Environment was held at Stockholm in 1972, there has been a growing emphasis within the international community to assess the environmental impact of new areas of technology which cut across national boundaries and to provide rights of private parties to participate in the policy debate. Evidence of this fact was noted in recommendations made by various international organizations, for example, the OECD's Council of Ministers in November 1974 enacted a principle of the "right to equal hearing," whereby individuals have the right to become involved in the investigative procedures of national jurisdictions in cases involving transfrontier pollution. The European Economic Community was also mentioned as they are presently studying the possibility of a Community-wide environmental impact assessment procedure. There are indications of national interest in the legislatures of Holland, United Kingdom, West Germany, Canada, and New Zealand, which currently have, or are in the process of formalizing, procedures dealing with environmental assessment.

The recent growth and apparent evolution of new rules, norms, and codes of conduct concerning assessment procedures which relate to new areas of technology suggest that although the full commercialization of OTEC is some 20 years off, one must be aware of this new consciousness and proceed accordingly.

Research Suggestions

There were several recommendations concerning possible research which could be undertaken at this time to study the best action to be taken in response to the evolving international interest in establishing new procedures for an international system of environmental assessment. The Council on Environmental Quality has been actively promoting this need for the future concern of procedures, and recently suggested that the present procedures which exist within the United States system of NEPA could be internationalized, whereby potentially affected countries or neighboring states would be asked to participate, submit their comments, which would then be circulated and integrated as part of the assessment process.

There were several suggestions made concerning the organization responsible for overseeing such a large-scale assessment procedure. The United Nations Environmental Program, while presently responsible for establishing standards related to the environment as well as an international information system with a consultative and notification procedure, was considered as a likely candidate. Another procedure might be the expansion of the bilateral agreement mechanism to include joint impact statements. Still another alternative was the possibility of nongovernmental international organizations, such as the International Council of Scientific Unions (ICSU), which could offer more of an independent and unbiased panel of experts to assess the risks and establish necessary guidelines. Finally, there was a recommendation to utilize the Engineering Committee on Oceanic Engineering (ECOR), which has a definite interest in OTEC. ECOR is represented in the United States through the National Academy of Engineering's Marine Board.

Long-range research suggestions included a study of possible climatic effects, so that the views of interested parties could be taken into account, and a consultative procedure established, as suggested by one of the Stockholm recommendations.

Report of Working Group IV: Economic and Financial Aspects of OTEC

Gerald Kent Fisher, Editor

The discussions of the Working Group on the economic and financial aspects of developing and deploying OTEC plants dealt with several disparate analytical problems associated with the changing demand and supply patterns of energy production and consumption, based on the papers by Messrs. Stern and Riggs. The analysis of whether OTEC is an economically feasible, alternative energy source obviously requires the making of several assumptions about the technology, future energy demands, the stability or instability of foreign energy supplies, costs of competing energy sources, the availability of capital, and the like; these same assumptions also color the commercial attractiveness of OTEC regarding investment analysis. In this context, Carlos Stern's chapter on economic issues attempted to outline the potential competitiveness of an OTEC system within current expectations regarding available energy sources. John Rigg's chapter on financing examined various techniques of financing OTEC development, construction, and use on the assumptions that the technological problems will be met and that OTEC thus could be part of a future energy supply mix.

The Working Group's discussions focused on two central issues that could critically affect the future position of OTEC as an energy supplier. First, the assumptions underlying economic analyses of energy issues are rapidly changing. Traditional techniques of cost and demand extrapolations are, at best, problematic. Externalities, such as environmental costs and the costs of dependence on foreign sources, must be, at least, mentioned. For OTEC in particular, the young technology makes cost-benefit analyses quite assumption-laden. Second, whereas the "source" (the ocean's thermal gradient) is "free," the OTEC system is a very capital-intensive technique of providing an alternative source of energy. This characteristic requires, if plants are to be deployed in sufficient numbers, a certain political priority applied to OTEC at a time when capital markets seem to be experiencing sharp demand increases. These two issue areas set the parameters within which the Working Group could frame its conclusions and recommendations regarding this type of solar energy converter.

The Working Group's conclusions and recommendations regarding the economic and financial aspects of OTEC can best be summarized if we look at OTEC's development in three particular stages, i.e., the research and development level, the interim stage (limited operation), and the mature system level.

Research and Development Stage

Although industry, such as Lockheed and TRW, has a demonstrated interest in the technology, government support of technology-related R&D is obviously essential. More sophisticated cost projections will be possible if the technical uncertainties of some of the working parts, e.g., the heat exchanger, the mooring, the transmission cable, are reduced through further testing and demonstration. Better information of this type will allow the continuous evaluation and reevaluation of OTEC economics vis-à-vis competing energy forms, both fossil and nonfossil in nature.

Interim Stage

At this juncture, the limited operational deployment of OTEC plants, a settled legal framework will be necessary if the financing of OTEC devices is to become at least partially private. Depending on technology and alternative source assessments, OTEC could be made as artificially competitive *as necessary* by governmental policies. Since the capital involved is so substantial, it was thought that statutory clarifications regarding the legal framework within which OTECs could be financed would have to be made, e.g., the potential applicability of the 1920 Ship Mortgage Act, the possible governmental guarantees of mortgage bonds under Title XI of the Merchant Marine Act of 1936 using existing MarAd mechanisms, or the use of lease financing if tax questions become settled. It was generally agreed that sophisticated private financing techniques, such as those used by oil companies, shipping firms, and utilities, could be applied to OTEC if it proves itself at this level.

Mature System

If OTEC proves "seaworthy" and competitive with other forms of energy production, the long-term question becomes one of public policy: What is the desired mix of public/private ownership of an energy source, or of governmental preferences and incentives to private markets regarding OTEC? Of course, OTEC could prove so commercially attractive that it captures a market on its own merits, but a realistic analysis of OTEC's economic feasibility and financial attractiveness must be based on what the then current alternatives are if it is to be placed in the proper policy perspective, and the allocation of resources to that policy must reflect this analysis.

Glossary

BTU British thermal unit
°C. degrees Centigrade
continental shelf submerged continuation of continental land mass over which adjacent coastal state has exclusive rights to non-living resources
CZMA Coastal Zone Management Act of 1972
DOD Department of Defense
DOMES Deep Ocean Mining Environmental Study (Department of Commerce)
DOT Department of Transportation
DWT dead weight tons
economic resource zone proposed zone of ocean space in which the adjacent coastal state would possess exclusive rights with respect to living and non-living ocean resources; usually limited at 200 mi from the coast
ERDA Energy Research and Development Administration
°F. degrees Fahrenheit
FHWA Federal Highway Administration
ft^2 square feet
ft^3 cubic feet
FWPCA Federal Water Pollution Control Act
high seas waters beyond the seaward extent of the territorial sea which are not subject to exclusive appropriation by any nation
hr hour
IEA International Energy Agency
IHO International Hydrographic Organization
IMCO Inter-governmental Maritime Consultative Organization
IRC Internal Revenue Code
IRS Internal Revenue Service
ISNT Informal Single Negotiating Text
kilo- times 1,000
kw (KW) kilowatt
LDC less developed country
LNG liquefied natural gas
LOS law of the sea
MarAd U.S. Maritime Administration (Department of Commerce)
mega- times 1,000,000
mw (MW) megawatt
NEPA National Environmental Policy Act (U.S.)
OCS outer continental shelf
OCSLA Outer Continental Shelf Lands Act
OPEC Organization of Petroleum Exporting Countries
OTEC ocean thermal energy conversion
seabed (deep seabed) ocean floor beyond limit of either the continental shelf or the economic resources zone

SEC Federal Securities and Exchange Commission

ship mortgage mortgage on a vessel under U.S. federal or foreign law

SNT Informal Single Negotiating Text

ΔT° thermal differential

take-or-pay contracts agreements under which a corporation or other person or entity absolutely agrees to purchase an agreed amount of a product at an agreed price over an agreed time period, regardless of whether such product is needed by the purchaser, payment therefor is received, or other conditions are fulfilled; a common basis for financings

territorial sea narrow band of coastal waters over which the adjacent coastal nation has virtually absolute sovereignty

Title XI Title XI of the federal Merchant Marine Act, 1936, as amended, relating to federal government guarantees of financings of U.S.-flag vessels owned by U.S. persons

UNCLOS III Third United Nations Conference on the Law of the Sea

UNEP United Nations Environment Program

VLCC Very Large Crude (oil) Carriers, or vessels, sometimes also referred to as "supertankers"

Index

Index

Members of the American Society of International Law Panel on Ocean Thermal Conversion

Bennett Boskey, Attorney, Washington, D.C.

Robert Cohen, Division of Solar Energy, Energy Research and Development Administration, Washington, D.C.

C.R. Hallberg, Office of the Chief Counsel, Coast Guard Headquarters, Washington, D.C.

James C. Higgins, Jr., Offshore Power Systems, Jacksonville, Florida.

Ann L. Hollick, Woodrow Wilson Center for Scholars, Washington, D.C.

H. Gary Knight, Louisiana State University Law Center, Baton Rouge, Louisiana

Arthur Konopka, National Science Foundation, Washington, D.C.

J.D. Nyhart, Sloan School of Management and Department of Ocean Engineering, Massachusetts Institute of Technology, Cambridge, Massachusetts.

John H. Riggs, Jr., White and Case, Brussels, Belgium.

Warren M. Rohsenow, Department of Mechanical Engineering, Massachusetts Institute of Technology, Cambridge, Massachusetts.

Herman E. Sheets University of Rhode Island.

Robert E. Stein, North American Office of the International Institute for Environment and Development in Washington, D.C., and American University Washington College of Law.

Carlos D. Stern, University of Connecticut. U.S. Geological Survey, Reston, Virginia.

Thomas B. Stoel, Jr., Natural Resources Defense Council, Washington, D.C.

Byron Washom, Department of Ocean Engineering of Massachusetts Institute of Technology, Cambridge, Massachusetts

Norman A. Wulf, National Science Foundation, Washington, D.C.

Guests of the Panel

Robert Douglass, TRW Systems, Inc., Redondo Beach, California.

Frederick E. Naef, Lockheed Missiles & Space Co., Washington, D.C.

Ralph G. Eldridge, The Mitre Corporation, McLean, Virginia.

L. Manning Muntzing, Attorney, Washington, D.C.

James P. Walsh, Attorney, Senate Commerce Committee, Washington, D.C.

John Tepe, Attorney, Washington, D.C.

About the Contributors

C.R. Hallberg holds the rank of Captain, U.S. Coast Guard, and is with the Office of the Chief Counsel, Headquarters, Washington, D.C.

James C. Higgins, Jr. is counsel of Offshore Power Systems, Jacksonville, Florida.

Ann L. Hollick is a fellow of the Woodrow Wilson Center for Scholars, Washington, D.C.

John H. Riggs, Jr. is an attorney with White and Case in Brussels, Belgium.

Herman E. Sheets is chairman of the Department of Ocean Engineering, University of Rhode Island.

Carlos D. Stern is assistant professor, College of Agriculture and Natural Resources, University of Connecticut. Presently with the Office of the Director, Program Analysis, U.S. Geological Survey, Reston, Virginia.

Thomas B. Stoel, Jr. is a staff attorney with the Natural Resources Defense Council, Washington, D.C.

Byron Washom is a special graduate student, Department of Ocean Engineering of Massachusetts Institute of Technology. He is on leave from the University of Southern California, formerly cochairman of the Ocean Economic Potential Committee, Marine Technological Society.

About the Editors

H. Gary Knight is Campanile Charities Professor of Marine Resources Law, Louisiana State University Law Center.

J.D. Nyhart is associate professor of management, Sloan School of Management and Department of Ocean Engineering; Coordinator of Law-related Studies, Massachusetts Institute of Technology.

Robert E. Stein is director of the North American Office of the International Institute for Environment and Development in Washington, D.C. and an adjunct professor of environmental law, American University Washington College of Law.

About the Society

The American Society of International Law was organized in 1906 and incorporated by special Act of Congress in 1950. Its objects are "to foster the study of international law and to promote the establishment and maintenance of international relations on the basis of law and justice."

Concerned with problems of international order and the legal framework for international relations for more than a half century, the Society serves as a meeting place, forum and collegial research center for scholars, officials, practicing lawyers, students, and others. The Society is hospitable to all viewpoints in its meetings and in its publications. Those publications include the leading periodicals, *The American Journal of International Law* and *International Legal Materials.* In addition, the Society publishes books, reports, and the occasional papers series, *Studies in Transnational Legal Policy*, produced by an extensive Research and Study Program under the supervision of its Board of Review and Development.

The Society's membership, which exceeds 5000, is drawn from some 100 countries. Membership is open to all, whatever their nationality or profession.